Catalytic, Photocatalytic and Electrocatalytic Processes for the Valorisation of CO_2

Catalytic, Photocatalytic and Electrocatalytic Processes for the Valorisation of CO$_2$

Special Issue Editors

Ilenia Rossetti
Gianguido Ramis

MDPI • Basel • Beijing • Wuhan • Barcelona • Belgrade

MDPI

Special Issue Editors

Ilenia Rossetti
Università degli Studi di Milano
Italy

Gianguido Ramis
Università degli Studi di Genova and INSTM Unit Genova
Italy

Editorial Office
MDPI
St. Alban-Anlage 66
4052 Basel, Switzerland

This is a reprint of articles from the Special Issue published online in the open access journal *Catalysts* (ISSN 2073-4344) from 2017 to 2019 (available at: https://www.mdpi.com/journal/catalysts/special_issues/valorisation_CO2).

For citation purposes, cite each article independently as indicated on the article page online and as indicated below:

LastName, A.A.; LastName, B.B.; LastName, C.C. Article Title. *Journal Name* **Year**, *Article Number*, Page Range.

ISBN 978-3-03921-778-6 (Pbk)
ISBN 978-3-03921-779-3 (PDF)

Contents

About the Special Issue Editors

Ilenia Rossetti (Prof.) Industrial chemist and chemical engineer, she is now Associate Professor of Chemical Plants at Università degli Studi di Milano, Italy. Her main research interests are the development of catalytic processes, starting from material design to kinetic modelling, to process design and simulation. She is the author of three books, 25 chapters/reviews, and 125 papers in international journals with an impact factor, and she has guest edited six Special Issues.

Gianguido Ramis (Prof.) An industrial chemist, he is now Associate Professor of Chemistry for Engineering at Università degli Studi di Genova, Italy. His main interests of research are the development of catalytic materials, from their formulation to their characterization through advanced spectroscopic techniques. He is author of ca. 160 papers in international journals with an impact factor, and he has guest edited three Special Issues.

catalysts

MDPI

Editorial

Catalytic, Photocatalytic, and Electrocatalytic Processes for the Valorization of CO_2

Ilenia Rossetti [1],* and Gianguido Ramis [2]

[1] Chemical Plants and Industrial Chemistry Group, Dip. Chimica, Università degli Studi di Milano, CNR-ISTM and INSTM Unit Milano-Università, via C. Golgi 19, 20133 Milan, Italy

[2] Dip. Ing. Chimica, Civile ed Ambientale, Università degli Studi di Genova and INSTM Unit Genova, via all'Opera Pia 15A, 16145 Genoa, Italy; gianguidoramis@unige.it

* Correspondence: ilenia.rossetti@unimi.it; Tel.: +39-02-503-14059; Fax: +39-02-503-14300

Received: 2 September 2019; Accepted: 4 September 2019; Published: 12 September 2019

Worldwide yearly CO_2 emissions reached 36 Gt in 2014, whereas they amounted to ca. 22 Gt in 1990 [1]. It represents the most abundant greenhouse gas in the atmosphere, 65% of which is derived from direct emissions (combustion and industrial processes), with an additional 11% from the change in land use and forestry [2].

An historic agreement to fight against climate change and move toward a low-carbon, resilient, and sustainable future was agreed by 195 nations in Paris (December 2015). The goal is to keep the temperature rise for the century less than 2 °C and to further limit it to 1.5 °C higher than pre-industrial values. The Intergovernmental Panel on Climate Change [3,4] evidenced an average increase of ca. 0.6 °C on Earth during the 20th century, with an almost exponential rise in the last decade.

Different strategies may be put into place to limit CO_2 output into the atmosphere, e.g., a more efficient use of carbon-based fossil fuels, the use of carbon-less or carbon-free raw materials, and, ultimately, CO_2 capture technologies. However, to be fully effective, carbon dioxide sequestration must be followed by its efficient conversion into useful new materials. This circular approach is virtuous, as it valorizes waste as a new regenerated feedstock, thus limiting the consumption of new sources. In this light, the research has turned from the simpler concept of carbon capture and storage (CCS) to carbon capture and conversion (CCC) or utilization (CCU) [5,6]. This approach moves toward a circular economy approach, perfectly matching the EU and international policies and initiatives.

The first capture step is common: The efficient removal of CO_2 from different point sources, such as the treatment of industrial flue gases, typically in stationary combustion plants, by a separation process, prior to the release of combustion exhausts into the atmosphere. It is much harder to imagine a sequestration method directly by absorption from the atmosphere, as the CO_2 concentration is too low to guarantee a sufficient driving force for its separation.

Several techniques are available for the capture of CO_2 from flue gas [7].

The main approaches are based on:

i) absorption, either chemical (with ammonia or amines) or physical (Rectisol, Selexol, or Fluor processes), where the selection of the solvent and optimization of the process are key for success [8,9];

ii) adsorption, usually including the regeneration of the adsorbent through pressure swing or temperature swing adsorption (PSA or TSA) [10];

iii) cryogenic separation;

iv) membrane separation (polymeric or ceramic materials);

v) hybrid technologies.

All of these processes present advantages and disadvantages, which have been very effectively reviewed in a recent paper [11], but the main technology for CO_2 capture is absorption with a liquid solvent (usually alcanolamines [9]) or adsorption in a PSA unit [10].

Once separated, former CCS approaches included planning the confinement of CO_2 into depleted oil and gas wells, deep oceans, and aquifers, or to use it as a fluid for fuel extraction (e.g., its injection in geological formations and the subsequent recovery of fuel products with several techniques, such as enhanced oil recovery, enhanced coal bed methane recovery, enhanced gas recovery). However, as said, much more effective approaches are needed to try to recover valuable products from its conversion through a circular economy vision [12].

In addition, in this case, different strategies are under development, leading to various upgraded products. Some conversion routes are oriented toward the regeneration of fuels, which can, in general, be defined as $C_xH_yO_z$. The energy required for CO_2 reduction increases more and more while decreasing z and increasing x, while the stored energy in the regenerated fuel increases accordingly. This approach is usually convenient when inexpensive renewable energy is available. In addition, it is a circular path, as the main CO_2 emissions come from combustion, and fuels are regenerated to be used in the same market. Provided a sufficiently efficient and cost-effective technology for fuel regeneration from CO_2 can be developed, the potential market for regenerated fuels matches the huge CO_2 emission volume. The opposite holds for rival pathways that tend to valorize CO_2 in fine chemicals or value-added products. If, on the one hand, the remunerability of the product is much higher, and can economically support the development of the technology, the size of the potential market of these fine chemicals can hardly match the volume of CO_2 emissions, needing the development of a network of parallel CCU technologies.

Among the different pathways, biological methods have been developed, either directly producing reduced products [13] (i.e., carbonic anhydrase, hydrogenation of CO_2 to formate, reduction of CO_2 to methane, CO_2 conversion into methanol by enzyme cascade) or storing CO_2 in biomass (e.g., algae) [14], to be subsequently upgraded.

The electrochemical reduction of CO_2 can be effectively exploited through available renewable electricity. For instance, CO_2 can be electrochemically reduced to formic acid derivatives that can subsequently be converted into useful monomers, such as glycolic acid and oxalic acid, to be employed as building blocks for polyesters [15]. The potential of CO_2 electroreduction to methane has been recently reviewed by Zhang et al. [16].

Photocatalytic reduction allows the production of a wide spectrum of products, such as HCOOH, HCHO, CH_3OH, or CH_4 [17–23], and can be effectively used for the exploitation of solar energy, provided visible responsive materials can be developed for this application [24].

Furthermore, the catalytic reduction of CO_2 has been proposed through the Sabatier reaction [25]. The methane produced in this reaction has great potential for application, but the application of this technology relies on the availability of renewable and inexpensive H_2. A different approach is the methanation of CO_2 through biochemical approaches.

This Special Issue

Different options of CO_2 valorization have been discussed in this special issue, showing interesting examples of the catalytic science impact in this important field. At first, the potential for the catalytic methanation of CO_2 was reviewed by Manzoli and Bonelli [26], who discussed the importance of catalyst design to ensure efficient performance in the (photo)catalytic hydrogenation of CO_2. Different enabling technologies assisting catalyst synthesis (microwaves, ultrasound, mechanochemical synthesis) allow tailored properties for the selected materials to be obtained, which, in turn, ensures suitable catalyst performance.

The methanation of CO_2 was studied under unsteady conditions to evidence the structural dynamics of the Ni/Al_2O_3 catalyst [27]. Different operando characterization techniques were used, such as Quick-scanning X-ray Absorption Spectroscopy/Extended X-ray Absorption Fine Structure

(XAS/QEXAFS) considering stops in the H_2 supply during the reaction while feeding oxidizing impurities to simulate the technical purity of CO_2.

Methanation has also been investigated through sorption-enhancement [28], which allows faster production of pure methane thanks to the application of Le Châtelier's principle. The long-term stability of a catalyst, constituted by Ni nanoparticles supported on zeolite 5A, has been examined and showed to be satisfactory thanks to milder operating conditions of the sorption-enhanced process with respect to the conventional one. A degradation mechanism specific to the sorption catalysis was derived on the basis of cyclic methanation/drying periods and was related to the water diffusion kinetics in the zeolite. The latter step is rate controlling during both methanation and drying, so this point is kinetically critical.

An example of the electro-reduction of CO_2 was proposed by Castelo-Quibén et al. [29], obtaining C1 to C4 hydrocarbons, as an efficient strategy for C–C coupling. The electroactive materials were composite metal–carbon–carbon nanofibers synthesized using urban plastic residues through catalytic pyrolysis. Selectivity was tunable by changing the metal.

The photoreduction of CO_2 was investigated under unconventional operating conditions, i.e., at high pressure and high temperature [21], using different TiO_2-based catalysts and investigating the effect of different reaction parameters. A significant productivity of liquid-phase products (HCOOH and HCHO) was achieved. The selectivity to different products was tuned based on pH, reaction time, and through the addition of Au nanoparticles as a co-catalyst.

Finally, the synthesis of dimethylcarbonate (DMC) was investigated by Han et al. [30,31]. Different alkali metals were added to Cu-Ni/diatomite catalysts to synthesize DMC from CO_2 and methanol, thanks to their strong electron-donating ability. Cs_2O was effective, leading to a ca. 10% methanol conversion with a ca. 86% selectivity to DMC [31]. Furthermore, the same authors investigated the effect of dehydration using 3A molecular sieves to shift the equilibrium and to improve the stability of a K_2O-promoted Cu–Ni catalyst [30]. An improved yield of DMC by 13% was obtained with respect to the undehydrated base case and stable performance for 22 h.

Author Contributions: I.R. and G.R. contributed equally to write this editorial.

Funding: This research received no external funding

Conflicts of Interest: The authors declare no conflict of interest

References

1. Available online: https://data.worldbank.org/indicator/EN.ATM.CO2E.KT?end=2014&start=1990 (accessed on 30 August 2019).
2. Available online: https://www.epa.gov/ghgemissions/global-greenhouse-gas-emissions-data (accessed on 30 August 2019).
3. Livingstone, J.E.; Lovbrand, E.; Olsson, J.A. From climates multiple to climate singular: Maintaining policy-relevance in the IPCC synthesis report. *Environ. Sci. Policy* **2018**, *90*, 83–90. [CrossRef]
4. Trainer, T. A critical analysis of the 2014 IPCC report on capital cost of mitigation and of renewable energy. *Energy Policy* **2017**, *104*, 214–220. [CrossRef]
5. Yaashikaa, P.R.; Senthil Kumar, P.; Varjani, S.J.; Saravanan, A. A review on photochemical, biochemical and electrochemical transformation of CO_2 into value-added products. *J. CO_2 Util.* **2019**, *33*, 131–147. [CrossRef]
6. Marocco Stuardi, F.; MacPherson, F.; Leclaire, J. Integrated CO_2 capture and utilization: A priority research direction. *Curr. Opin. Green Sustain. Chem.* **2019**, *16*, 71–76. [CrossRef]
7. Usubharatana, P.; MCMartin, D.; Veawab, A.; Tontiwachwuthikul, P. Photocatalytic process for CO_2 emission reduction from industrial flue gas streams. *Ind. Eng. Chem. Res.* **2006**, *45*, 2558–2568. [CrossRef]
8. Borhani, T.N.; Wang, M. Role of solvents in CO_2 capture processes: The review of selection and design methods. *Renew. Sustain. Energy Rev.* **2019**, *114*, 109299. [CrossRef]
9. de Guido, G.; Compagnoni, M.; Pellegrini, L.A.; Rossetti, I. Mature versus emerging technologies for CO_2 capture in power plants: Key open issues in post-combustion amine scrubbing and in chemical looping combustion. *Front. Chem. Sci. Eng.* **2018**, *12*, 315–325. [CrossRef]

10. Zhu, X.; Li, S.; Shi, Y.; Cai, N. Recent advances in elevated-temperature pressure swing adsorption for carbon capture and hydrogen production. *Prog. Energy Combust. Sci.* **2019**, *75*, 100784. [CrossRef]

11. Asif, M.; Suleman, M.; Haq, I.; Jamal, S.A. Post-combustion CO_2 capture with chemical absorption and hybrid system: Current status and challenges. *Greenh. Gases Sci. Technol.* **2018**, *8*, 998–1031. [CrossRef]

12. Rafiee, A.; Rajab Khalilpour, K.; Milani, D.; Panahi, M. Trends in CO_2 conversion and utilization: A review from process systems perspective. *J. Environ. Chem. Eng.* **2018**, *6*, 5771–5794. [CrossRef]

13. Bhatia, S.K.; Bhatia, R.K.; Jeon, J.M.; Kumar, G.; Yang, Y.H. Carbon dioxide capture and bioenergy production using biological system—A review. *Renew. Sustain. Energy Rev.* **2019**, *110*, 143–158. [CrossRef]

14. Cheng, J.; Zhu, Y.; Zhang, Z.; Yang, W. Modification and improvement of microalgae strains for strengthening CO_2 fixation from coal-fired flue gas in power plants. *Bioresour. Technol.* **2019**, *291*, 121850. [CrossRef] [PubMed]

15. Murcia Valderrama, M.A.; van Putten, R.-J.; Gruter, G.-J.M. The potential of oxalic—And glycolic acid based polyesters (review). Towards CO_2 as a feedstock (Carbon Capture and Utilization—CCU). *Eur. Polym. J.* **2019**, *119*, 445–468. [CrossRef]

16. Zhang, Z.; Song, Y.; Zheng, S.; Zhen, G.; Lu, X.; Takuro, K.; Xu, K.; Bakonyi, P. Electro-conversion of carbon dioxide (CO_2) to low-carbon methane by bioelectromethanogenesis process in microbial electrolysis cells: The current status and future perspective. *Bioresour. Technol.* **2019**, *279*, 339–349. [CrossRef] [PubMed]

17. Kim, J.; Kwon, E.E. Photoconversion of carbon dioxide into fuels using semiconductors. *J. CO_2 Util.* **2019**, *33*, 72–82. [CrossRef]

18. Bahdori, E.; Tripodi, A.; Villa, A.; Pirola, C.; Prati, L.; Ramis, G.; Dimitratos, N.; Wang, D.; Rossetti, I. High pressure CO_2 photoreduction using Au/TiO_2: Unravelling the effect of the co-catalyst and of the titania polymorph. *Catal. Sci. Technol.* **2019**, *9*, 2253–2265. [CrossRef]

19. Rossetti, I.; Villa, A.; Pirola, C.; Prati, L.; Ramis, G. A novel high-pressure photoreactor for CO_2 photoconversion to fuels. *RSC Adv.* **2014**, *4*, 28883–28885. [CrossRef]

20. Galli, F.; Compagnoni, M.; Vitali, D.; Pirola, C.; Bianchi, C.L.; Villa, A.; Prati, L.; Rossetti, I. CO_2 photoreduction at high pressure to both gas and liquid products over titanium dioxide. *Appl. Catal. B Environ.* **2017**, *200*, 386–391. [CrossRef]

21. Bahadori, E.; Tripodi, A.; Villa, A.; Pirola, C.; Prati, L.; Ramis, G.; Rossetti, I. High Pressure Photoreduction of CO_2: Effect of Catalyst Formulation, Hole Scavenger Addition and Operating Conditions. *Catalysts* **2018**, *8*, 430. [CrossRef]

22. Rossetti, I.; Villa, A.; Compagnoni, M.; Prati, L.; Ramis, G.; Pirola, C.; Bianchi, C.L.; Wang, W.; Wang, D. CO_2 photoconversion to fuels under high pressure: Effect of TiO_2 phase and of unconventional reaction conditions. *Catal. Sci. Technol.* **2015**, *5*, 4481–4487. [CrossRef]

23. Compagnoni, M.; Ramis, G.; Freyria, F.S.; Armandi, M.; Bonelli, B.; Rossetti, I. Innovative photoreactors for unconventional photocatalytic processes: The photoreduction of CO_2 and the photo-oxidation of ammonia. *Rend. Lincei* **2017**, *28*, 151–158. [CrossRef]

24. Rossetti, I.; Bahadori, E.; Tripodi, A.; Villa, A.; Prati, L.; Ramis, G. Conceptual design and feasibility assessment of photoreactors for solar energy storage. *Sol. Energy* **2018**, *172*, 225–231. [CrossRef]

25. Navarro, J.C.; Centeno, M.A.; Laguna, O.H.; Odriozola, J.A. Policies and motivations for the CO_2 valorization through the sabatier reaction using structured catalysts. A review of the most recent advances. *Catalysts* **2018**, *8*, 578. [CrossRef]

26. Manzoli, M.; Bonelli, B. Microwave, ultrasound, and mechanochemistry: Unconventional tools that are used to obtain "smart" catalysts for CO_2 hydrogenation. *Catalysts* **2018**, *8*, 262. [CrossRef]

27. Mutz, B.; Gänzler, A.M.; Nachtegaal, M.; Müller, O.; Frahm, R.; Kleist, W.; Grunwaldt, J.D. Surface oxidation of supported Ni particles and its impact on the catalytic performance during dynamically operated methanation of CO_2. *Catalysts* **2017**, *7*, 279. [CrossRef]

28. Delmelle, R.; Terreni, J.; Remhof, A.; Heel, A.; Proost, J.; Borgschulte, A. Evolution of water diffusion in a sorption-enhanced methanation catalyst. *Catalysts* **2018**, *8*, 341. [CrossRef]

29. Castelo-Quibén, J.; Elmouwahidi, A.; Maldonado-Hódar, F.J.; Carrasco-Marín, F.; Pérez-Cadenas, A.F. Metal-carbon-CNF composites obtained by catalytic pyrolysis of urban plastic residues as electro-catalysts for the reduction of CO_2. *Catalysts* **2018**, *8*, 198. [CrossRef]

Catalysts **2019**, *9*, 765

30. Han, D.; Chen, Y.; Wang, S.; Xiao, M.; Lu, Y.; Meng, Y. Effect of in-situ dehydration on activity and stability of Cu–Ni–K2O/diatomite as catalyst for direct synthesis of dimethyl carbonate. *Catalysts* **2018**, *8*, 343. [CrossRef]
31. Han, D.; Chen, Y.; Wang, S.; Xiao, M.; Lu, Y.; Meng, Y. Effect of alkali-doping on the performance of diatomite supported cu-ni bimetal catalysts for direct synthesis of dimethyl carbonate. *Catalysts* **2018**, *8*, 302. [CrossRef]

catalysts

MDPI

Review

Microwave, Ultrasound, and Mechanochemistry: Unconventional Tools that Are Used to Obtain "Smart" Catalysts for CO_2 Hydrogenation

Maela Manzoli [1] and Barbara Bonelli [2,*

[1] Department of Drug Science and Technology, NIS and INSTM reference Centres,
    Università degli Studi di Torino, Via Pietro Giuria 9, 10125 Torino, Italy; maela.manzoli@unito.it
[2] Department of Applied Science and Technology and INSTM-Unit of Torino Politecnico,
    Corso Duca degli Abruzzi 24, 10129 Torino, Italy
* Correspondence: barbara.bonelli@polito.it; Tel.: +39-011-090-4719

Received: 6 April 2018; Accepted: 21 June 2018; Published: 28 June 2018

Abstract: The most recent progress obtained through the precise use of enabling technologies, namely microwave, ultrasound, and mechanochemistry, described in the literature for obtaining improved performance catalysts (and photocatalysts) for CO_2 hydrogenation, are reviewed. In particular, the main advantages (and drawbacks) found in using the proposed methodologies will be discussed and compared by focusing on catalyst design and optimization of clean and efficient (green) synthetic processes. The role of microwaves as a possible activation tool used to improve the reaction yield will also be considered.

Keywords: CO_2 hydrogenation; microwaves; ultrasound; mechanochemistry; catalyst preparation

1. Introduction

Atmospheric CO_2 concentration is growing continuously: in 2017, a "record concentration" of 413 ppm was registered (according to https://www.co2.earth/daily-co2) and a concentration of *ca.* 570 ppm is expected to be reached by the end of this century [1].

The mitigation of CO_2 (the most popular among atmospheric greenhouse gases) is currently a scientific, technological, social, and economic issue [1]. Several strategies have been proposed to limit/decrease CO_2 concentration in the atmosphere: besides lowering CO_2 emissions with the help of low-C fuels (and of appropriate policies from developed countries), there is a great spur of studies and technologies aimed at optimizing several processes of CO_2 capture & storage [2–10] and reutilization [11,12]. The first two steps, currently fairly developed, are indeed functional to CO_2 reutilization processes, which mostly need a concentrated and rather pure CO_2 stream apt to further chemical transformations.

The chemical processes for CO_2 reutilization are catalytic [13–15], photocatalytic [16–20], electrocatalytic [21–25], or photoelectrocatalytic [18,26]. As a whole, the chemical reduction of CO_2 is a challenge from both the industrial and the catalytic point of view [27], but it is also a hot scientific topic, since CO_2 is an ideal C1 feedstock for producing different types of chemicals and, ultimately, fuels. Unfortunately, its high thermodynamic stability limits the number of processes in which CO_2 can be utilized as a chemical feedstock (e.g., the synthesis of urea), and usually electrocatalytic processes are needed [28].

Most of the catalytic processes (like those reported in Scheme 1) imply the use of H_2, which is *per se* a costly reagent due to its scarce natural abundance (from which arises the issue of its production), along with safety and transportation issues. All this notwithstanding, the products of CO_2 reduction are ultimately fuels, raw materials, and/or chemical intermediates [1].

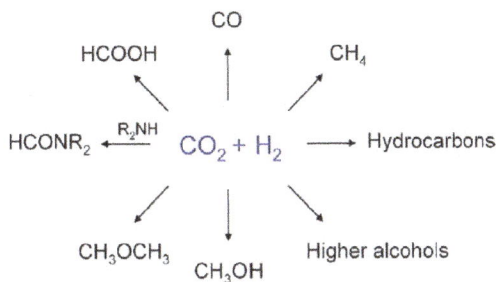

Scheme 1. Some products of CO_2 hydrogenation. Reproduced from Ref. [1] with permission of The Royal Society of Chemistry.

Though some homogenous catalysts are highly active and selective in several CO_2 reduction processes [29], they are hardly applicable on an industrial scale, and therefore there is a great interest towards the development of efficient heterogeneous catalysts and/or the immobilization of homogeneous ones [30]. For the industrial application of CO_2 hydrogenation processes, heterogeneous catalysts (e.g., Fe-, Cu-, and Ni-based solid catalysts [31]) offer some practical advantages compared to homogeneous ones: solids with large specific surface area, small particles of active phase, and fair metal dispersion are usually more active and selective, and also more stable and characterized by longer life. However, they often provide low yields and poor selectivity, mainly due to fast kinetics of C–H bond formation [1].

In order to allow industrial applications of heterogeneous catalysis, green processes should be developed, which often require efficient (nano)catalysts that are able to selectively transform CO_2 into added value molecules [27]. Depending on the reaction adopted for CO_2 reduction, different active phases (mostly Ni, but also more noble metals [31]) have to be supported on different oxides (γ-Al_2O_3 being the most used one [31]) that have to be stable at high temperatures, provide a certain amount of acid/base sites, and stabilize the catalytically active metal particles, etc.

Whatever the type of CO_2 hydrogenation process, active, selective, and stable heterogeneous catalysts are preferable in view of an actual industrial application [1]. They should also be based on earth-abundant and non-toxic elements and obtained by green synthesis methods: in one word, they should be "smart". Though the outputs of CO_2 reduction are diverse [1], "smart" catalysts have to meet some requirements that can be obtained by optimizing their preparation strategy, for instance. The physico-chemical properties of a heterogeneous catalyst are determined and controlled by its preparation method: therefore, catalysts prepared by different means may have markedly different features and, thus, different catalytic performances [32,33].

In order to allow an actual development of feasible industrial CO_2 hydrogenation processes, new catalysts should be obtained by methods that are able to control and tailor the physico-chemical properties in order to improve the catalyst performance [32]. Moreover, a deeper understanding of the reaction mechanisms with the help of both experiments and molecular simulations is needed [1], especially as far as the first step of CO_2 reduction, i.e., its adsorption at the catalyst surface, is concerned. Notwithstanding the plethora of works concerning CO_2 reduction, a thorough understanding of the mechanisms is still lacking and/or debated [18]. As far as heterogeneous catalysts are concerned, there is a general consensus about the idea that the active site is provided by the primary catalyst (e.g., the metal particle) acting synergistically with the support (and/or the promoter) [32]. Evidence of such synergy is provided in several works, but the actual mechanism is still unclear, even for the synthesis of methanoic acid (HCOOH), which is only the first step of CO_2 hydrogenation [1].

From both industrial and scientific points of view, the great interest in heterogeneous catalysts based on abundant and low cost metals like Ni, Cu, and Fe is frustrated by their poor activity in mild reaction conditions, in that an extreme dispersion of metal particles and a strong interaction with the support should be attained [31].

As mentioned before, from an environmental point of view, catalysts should be obtained by green synthesis methods [27]. In our opinion, the use of microwaves, ultrasound, or mechanochemistry represents promising and simple approaches, which are alternatives to conventionally adopted synthetic protocols, to obtain novel catalysts with improved activity and selectivity, in contrast to conventional methods, which usually encompass multi-step processes, heating, and/or the addition of expensive and/or hazardous chemicals. The development of such techniques has increased significantly in recent years, with most reports published over the past 5–10 years (*vide infra*).

The aim of the present review is to supply a brief (but exhaustive) description of the most recent advances in the synthesis of solid catalysts (and photocatalysts) by microwave (MW), ultrasound (US), and mechanochemistry (MC) to be compared to conventional methods.

2. Microwaves as a Tool to Improve Catalytic CO_2 Hydrogenation

2.1. General Principles of MW-Assisted Synthesis of (Nano)Materials

MW-assisted synthesis simply requires the presence of a MW reactor, simultaneously ensuring shorter reaction time, no contact between the radiation source and the reacting mixture, and a homogeneous heating of the latter [34,35].

Homogeneous heating is a crucial point, as thermal gradients affect both nucleation and growth steps, which finally determine the physico-chemical properties of the resulting nanomaterials. For this reason, MW-assisted synthesis was initially developed to prepare ceramic materials, for which it is crucial to avoid temperature gradients that may induce formation of fractures in the final product [36].

MW irradiation favors fast nucleation and crystalline growth, which can accelerate the formation of hierarchical structures, like the bismuth oxyiodide (BiOI) architectures shown in Figure 1 [37]. Hierarchical porous solids can be produced by MW-assisted synthesis in non-aqueous solvents, since organic solvents with high dielectric loss factor are required to ensure efficient MW absorption and rapid heating [37].

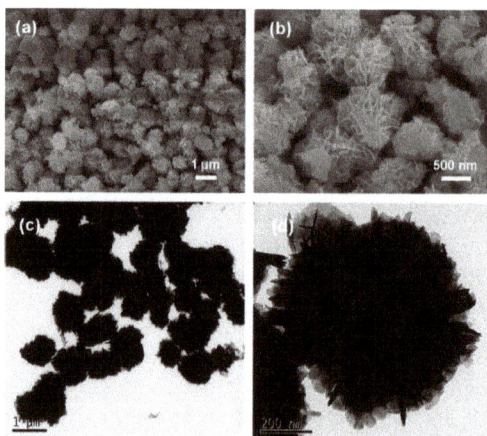

Figure 1. SEM (**a,b**) and TEM (**c,d**) images of BiOI flower-like spheres (dimensions in the 500 nm^{-1} μm range) made up by nanosheets as building blocks. Reprinted from *Chem. Eng. J.*, 235 Ai, L.; Zeng, Y.; Jiang, J., Hierarchical Porous BiOI Architectures: facile Microwave Nonaqueous Synthesis, Characterization, and Application in the Removal of Congo Red from Aqueous Solution, 331–339. Copyright (2014) with permission from Elsevier.

Additionally, MW heating can be easily coupled with several chemical methods including combustion synthesis (CS), hydrothermal (HT) synthesis, and sol-gel (SG). For instance, powders obtained by MW-assisted CS are more chemically and thermally stable than those obtained by "conventional" CS; moreover, smaller particles with a lower tendency to aggregation are obtained [36]. This occurs because MW irradiation uniformly heats the reagents mixture, whereas an electric oven heats a mixture from the exterior to the interior, finally creating a temperature gradient and leading to heterogeneous powders. Moreover, MW interacts with polar reagents, inducing dipole oscillations: the oscillating species collide more frequently and at higher energy, facilitating bond breakage and new bond formation. Consequently, the heat needed for the reaction is generated within the mixture with no need of external heat supply [34,35].

MW-assisted CS, for instance, allows obtaining sizeable amounts of phases (mainly oxides, nitrides, and carbides) that are stable at high temperatures and have nanometer size, highly specific surface area and defectivity, and mechanically stable pore structure [36]. MW is also known to favor the formation of given crystalline phases and can be efficiently used to drive the synthesis towards the desired products. All the above parameters are indeed important for catalytic applications.

2.2. MW-Assisted Syntheses of Heterogeneous Catalysts for CO_2 Reduction

Cai et al. [38] supported some Cu-based catalysts on ZnO and on Ga_2O_3 and tested them in the CO_2 hydrogenation to CH_3OH:

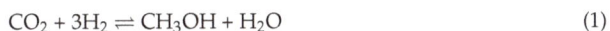

$$CO_2 + 3H_2 \rightleftharpoons CH_3OH + H_2O \tag{1}$$

The Cu-based catalysts were prepared by co-precipitation and impregnation, and the effect of MW on both preparation methods was studied: as a whole, the catalysts prepared by MW-assisted synthesis showed better Cu dispersion and enhanced catalytic activity in terms of both CO_2 conversion and CH_3OH yield, especially when a single-step MW-assisted method was adopted [38]. The catalysts were characterized by X-Ray Diffraction (XRD), H_2 temperature programmed reduction (TPR), N_2 adsorption/desorption isotherms at $-196\,°C$, CO_2 temperature-programmed desorption (TPD), adsorption calorimetry, and X-ray Photoemission Spectroscopy (XPS). Since all the samples are calcined at $350\,°C$, no significant changes in their surface area were observed, although the sample obtained by a single-step, MW-assisted method showed slightly larger values for both surface area and pore volume. The XRD patterns of the same sample revealed the likely presence of small crystallites of Cu_2O (whereas CuO mostly formed in the other samples) and the smallest size of ZnO particles. TPR showed the occurrence of a strong Cu/support interaction in the sample obtained by single-step, MW-assisted method, which, according to CO_2 TPD and CO_2 adsorption calorimetry, adsorbed a considerable amount of CO_2 with the largest average adsorption heat value (ca. $-20\,kJ\,mol^{-1}$). This latter result showed that the use of MW seemed to improve the support basicity, a feature that helps CO_2 adsorption, a crucial step during the catalytic process. The joint presence of well-dispersed Cu particles and strong basic sites at the support surface resulted in strong Cu/support synergy, leading to better catalytic performance. Besides strong Cu/support interactions, which took place with the catalysts prepared under MW irradiation, the authors observed that Cu particles were particularly active, probably because their surface was partially oxidized due to migration of Zn species from the support to the active metal (Cu) surface, which finally improved CO_2 adsorption. On the contrary, such synergistic effects were not observed with same composition catalysts that were prepared by conventional methods.

Huang et al. [39] studied the effect of MW heating during HT synthesis of $CuO-ZnO-ZrO_2$ catalysts for CO_2 hydrogenation to CH_3OH (Equation (1)), for which CuO-ZnO is usually considered the best catalyst and where ZrO_2 has the role of imparting high thermal and chemical stability under redox conditions. This type of catalyst is usually prepared by HT method, requiring long reaction times: the (expected) effect of MW irradiation was indeed the shortening of preparation time. Different temperatures (i.e., 80, 100, 120, and 150 °C) were adopted during the MW-assisted HT method to obtain the catalysts named CZZ80, CZZ100, CZZ120, and CZZ150 [39].

The catalyst obtained at 120 °C (CZZ120) showed the best catalytic properties in terms of both CO_2 conversion and selectivity to CH_3OH, becoming rather stable after 100 h reaction. XRD analysis allowed the calculation of the size of both CuO and Cu crystallites: the CZZ120 catalyst showed the smallest CuO and Cu crystallites before and after reduction, respectively, though Cu crystallites (ca. 18 nm) were larger than parent CuO crystallites (ca. 7.6 nm). The same catalyst was also characterized by the largest Cu area (S_{Cu}); the largest specific surface area and by a higher amount of CO_2 molecules adsorbed by relatively weak basic sites. CuO crystallites in CZZ120 were easily reduced, as the samples showed the lowest reduction temperature during H_2 TPR analysis. The CH_3OH yield was directly related to the surface of Cu particles (S_{Cu}), as shown in Figure 2. On the contrary, XPS analysis did not show significant differences in the type of surface Cu species in catalysts obtained at different temperatures. As a whole, an optimum temperature existed for the synthesis of $CuO-ZnO-ZrO_2$ catalysts [39]: Cu particle size, surface area, and basicity exhibited a volcano trend with a maximum at 120 °C.

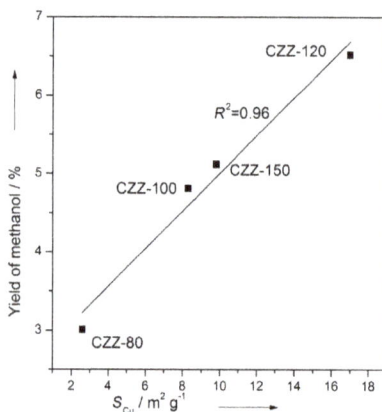

Figure 2. Linear relation between the CH_3OH % yield and the surface of Cu particles (S_{Cu}). Reprinted from Huang, C.; Mao, D.; Guo, X.; Yu, J., Microwave-Assisted Hydrothermal Synthesis of $CuO-ZnO-ZrO_2$ as Catalyst for Direct Synthesis of Methanol by Carbon Dioxide Hydrogenation. *Energy Technol.* Copyright (2017) with permission from Wiley.

CO_2 methanation (Equation (2)) is a process that allows transformation of CO_2 in a fuel through an exothermic reaction ($\Delta H^0 = -165$ kJ mol^{-1}), for which Ni/Al_2O_3 is the commercial catalyst [40]:

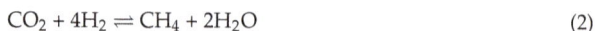

$$CO_2 + 4H_2 \rightleftharpoons CH_4 + 2H_2O \tag{2}$$

Besides Al_2O_3, other supports can be (mostly) SiO_2 and TiO_2, ZrO_2 or CeO_2. When other (more noble) metals like Ru, Pt, Pd, or Au are used, selectivity is lower with respect to Ni, as they do not produce exclusively CH_4, but also CH_3OH and CO (by Reversed Water Gas Shift, RWGS):

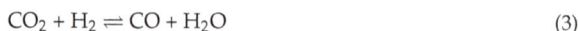

$$CO_2 + H_2 \rightleftharpoons CO + H_2O \tag{3}$$

As a whole, Ni dispersion is crucial in the catalytic activity towards CO_2 methanation [41–43], and therefore the catalyst synthesis method plays a prominent role [44–46]: according to Stangeland et al. [40], catalysts should be developed with improved activity at lower temperatures and high activity/stability up to ca. 500 °C. Conventional co-precipitation and impregnation methods may require several hours in order to obtain a fair Ni dispersion, whereas under MW irradiation the preparation time may decrease from hours to 15 min (this is one of the greatest advantages of using MW irradiation during co-precipitation and/or impregnation).

Moreover, MW-assisted synthesis of Ni/Al_2O_3 catalysts for CO_2 methanation [47] resulted in higher Ni dispersion, as well as higher basicity of the support with respect to Ni/Al_2O_3 catalysts obtained by impregnation. Physico-chemical characterization of the catalysts showed some relevant differences: for instance, Raman spectroscopy revealed a higher amount of NiO oxide in the sample prepared by impregnation, which also showed lower surface area ($\cong 143$ m^2 g^{-1} against 163 m^2 g^{-1}). The most important property was, however, the Ni dispersion: it was likely favored by the fact that MW helps the dispersion of inorganic salts on the support, as it can be clearly seen in Figure 3, showing the TEM analysis of two Ni/Al_2O_3 catalysts produced with and without MW assistance. The average size of Ni particles obtained by MW assisted synthesis was 10 nm and 16 nm in the sample prepared by impregnation. The sample obtained by MW-assisted synthesis also consumed ca. 25% more H_2 during TPD analysis.

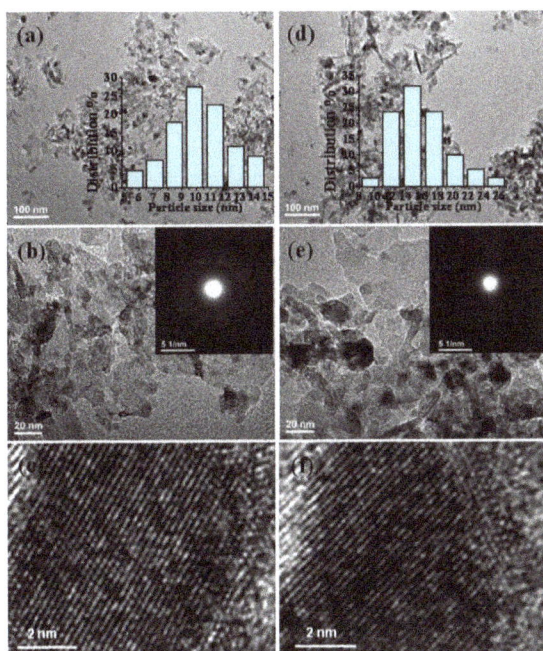

Figure 3. TEM, HRTEM, and SAED patterns of Ni/Al_2O_3 prepared under MW radiation (**a–c**) and without MW irradiation (**d–f**). Reprinted from *Int. J. Hydrogen Energy*, 42, Song, F.; Zhong, Q.; Yu, Y.; Shi, M.; Wu, Y.; Hu, J.; Song, Y., Obtaining Well-Dispersed Ni/Al_2O_3 Catalyst for CO_2 Methanation with a Microwave-Assisted Method, 4174–4183. Copyright (2017) with permission from Elsevier.

The better Ni dispersion obtained by MW joint with a stronger basicity of support (as determined by CO_2 TPD) likely led to enhanced CO_2 methanation activity at lower temperature. The corresponding CH_4 selectivity was also maintained over the whole temperature range (200–400 °C), and the catalyst was rather stable under prolonged (72 h) catalytic tests without appreciable agglomeration of Ni species [47].

MW can be used during catalyst preparation in different manners: for instance, the effect of MW drying vs. conventional drying in the preparation of Ni-exchanged zeolites for CO_2 methanation (2) has been recently studied [48]. The catalysts were prepared by incipient wetness impregnation of USY (Ultra Stabilized Y) zeolite with 5 wt % Ni. The effects of drying method after impregnation, calcination temperature, and pre-reduction temperature on the catalyst performance were evaluated. The drying method deeply affected the catalyst performance: in particular, MW-assisted drying induced remarkable changes on the type, location, and reducibility of Ni species within the zeolite structure and, simultaneously, led to effects on the support structural and textural properties and on the average size of Ni particles.

DR-UV-Vis spectroscopy showed that after conventional drying, most NiO species occurred at the outer surface of zeolite particles, whereas with MW-assisted drying, octahedral Ni^{2+} species formed if drying was performed in a closed container (when MW-assisted drying was performed in an open container, similar Ni species were obtained as they were during conventional drying). According to the authors, during MW heating in a close container, higher temperature and pressure led to formation of Ni^{2+} species in exchange position within the zeolite cavities, whereas if the container was left open during drying, NiO species mostly occurred at the outer surface of zeolite particles.

MW-assisted drying also deeply affected zeolite crystallinity: in a closed container, some amorphous phase formed (as detected by X-Ray powders diffraction), likely due to the high temperature reached within the container, whereas with MW-assisted drying in an open container, the zeolite crystallinity was preserved (as for conventional drying). Finally, MW-assisted drying led to less reducible Ni species, especially when octahedral Ni^{2+} species occurred. As observed with other Ni-based catalysts, MW-assisted drying led to formation of smaller Ni particles (in the 17–27 nm range). Unfortunately, the positive effects brought about by MW-assisted drying, like better Ni dispersion, were counterbalanced by the loss in zeolite crystallinity and, as a result, no appreciable improvement in the catalytic activity was obtained in terms of both CO_2 conversion and selectivity to CH_4. This was likely due to the fact that the support was a zeolite, for which the crystallinity of porous cavities is a crucial aspect [48].

The material thermal response to MW heating is a crucial aspect, as shown by Dong et al. [49], with Re promoted Ni-Mn/Al_2O_3 catalysts for partial methanation (Scheme 2), a process in which the amount of (noxious) CO is reduced by coupling CO methanation ($\Delta H^0 = -206$ kJ mol^{-1}):

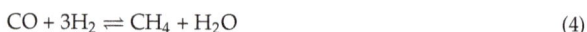

$$CO + 3H_2 \rightleftharpoons CH_4 + H_2O \tag{4}$$

to water gas shift (WGS, $\Delta H^0 = -41$ kJ mol^{-1})

$$CO + H_2O \rightleftharpoons CO_2 + H_2 \tag{5}$$

Scheme 2. Scheme of the processes occurring at the surface of Re-promoted Ni-Mn bimetallic particles on Al_2O_3 support. Reprinted from *Appl. Catal. A Gen.*, 552, Dong, X.; Jin, B.; Sun, Y.; Shi, K.; Yu, L., Re-Promoted Ni-Mn Bifunctional Catalysts Prepared by Microwave Heating for Partial Methanation Coupling with Water Gas Shift under Low H_2/CO Conditions. Copyright (2018) with permission from Elsevier.

Figure 4 reports heating curves under MW irradiation (1.0 kW output power) of the different reagents used to produce the catalysts: Al_2O_3 is a poor MW absorber, but when Re was firstly incorporated in γ-Al_2O_3 (Re-Al_2O_3), the materials showed heating curves with higher slope (curves 3 and 4 in Figure 4). Due to the presence of Re, both the Re-Al_2O_3 and Ni-Mn/Re-Al_2O_3 materials absorbed more MW with respect to pure γ-Al_2O_3. This had a deep impact on the reaction time, in that the synthesis temperature (ca. 400 °C) was reached shortly when Re-Al_2O_3 was used for the deposition of the bimetallic Ni-Mn catalyst starting from the respective nitrates.

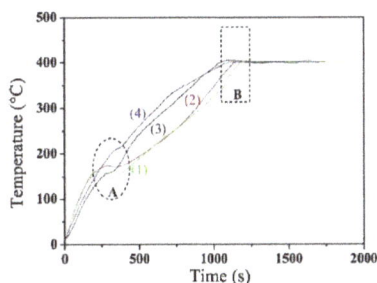

Figure 4. MW heating curves of different catalyst precursors: (**1**) Ni-Mn/Al_2O_3, (**2**) Ni-Mn-Re/Al_2O_3, (**3**) Re-Al_2O_3, and (**4**) Ni-Mn/Re-Al_2O_3. Reprinted from *Appl. Catal. A Gen.*, 552, Dong, X.; Jin, B.; Sun, Y.; Shi, K.; Yu, L., Re-Promoted Ni-Mn Bifunctional Catalysts Prepared by Microwave Heating for Partial Methanation Coupling with Water Gas Shift under Low H_2/CO Conditions. Copyright (2018) with permission from Elsevier.

Other phenomena occurring during MW heating can be inferred by a close inspection of the curves in Figure 4: when Re is absent, the temperature plateau reached at ca. 150 °C is ascribed to thermal balance between (Ni and Mn) nitrates decomposition and MW irradiation. The phenomenon does not occur in the heating curves of Re-added samples, likely because of the enhancement of thermal decomposition of nitrates due to the presence of Re. Simultaneously, MW heating favors the formation of hot spots that induce an effective surface dispersion of the metal. A side-effect of fast MW heating was observed, i.e., formation of AlO(OH) phase (as detected by X-ray diffraction). MW heating, however, favored the dispersion of the NiO phase, which showed broader X-ray diffraction peaks with respect to the catalyst obtained by conventional heating. A better dispersion of the Ni active phase was also observed after reduction, confirming the overall positive effect of MW. At variance with what was observed with Ni-USY zeolite, MW heated samples were more easily reducible, as the onset of NiO reduction shifted to lower temperatures. Another important point observed by the authors was that MW-heated catalysts led to lower formation of carbon deposits, an undesired phenomenon that can lead to catalyst deactivation [49].

The presence of Re, however, also affected some important structural and electronic properties of the catalysts, as shown by the characterization of their physico-chemical properties. For instance, H_2 TPR analysis showed a higher reducibility of β-type NiO (a type of NiO that has medium strong interaction with γ-Al_2O_3), indicating the facile formation of catalytically active Ni atoms. XPS analysis pointed out also a decrease of the binding energy of the Ni 2p line when Re was present, ascribed to the electron transfer from ReO_x species to Ni atoms. The increased electron cloud density of Ni atoms could indeed favor the breakage of the C-O bond, positively affecting the catalytic activity.

CO_2 reforming of methane (Equation (6)) is an endothermic reaction (ΔH^0 = +247 kJ mol^{-1}, also known as dry reforming) to produce syngas mixtures with low H_2:CO ratio that could be used for further Fischer-Tropsch synthesis (FTS) to produce light olefins:

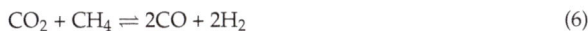

$$CO_2 + CH_4 \rightleftharpoons 2CO + 2H_2 \tag{6}$$

The reaction has not yet been industrially implemented, since there are no commercial catalysts that can operate without undergoing deactivation due to carbon deposition [50]. An alternative to Ni-based catalysts (and to noble metals like Ru, Rh, Pd, Ir, and Pt) is provided by Ni-containing oxides, i.e., either Ni perovskites with defined structure or solid solutions with La, Al, or Mg (spinel, fluorite, and perovskite) [51]. Ni perovskites with defined structure are promising catalysts, with respect to noble metals, due to their low cost, thermodynamic stability at relatively high temperatures, and catalytic activity. Barros et al. [51] prepared La_2NiO_4/α-Al_2O_3 catalysts for CO_2 reforming of CH_4 by MW-assisted CS, by using glycine or urea as fuel and metal nitrates as oxidizers. For comparison, a catalyst with similar chemical composition was prepared by impregnation of α-Al_2O_3 prepared by MW.

XRD diffraction showed the formation of $La_3Ni_2O_{6.92}$, La_2NiO_4, and $LaAl_xNi_{x-1}O_3$ crystalline structures independently of the preparation method. The sample prepared by impregnation was very active, but the crystalline structure was practically destroyed, with significant carbon deposition. The catalysts prepared by MW-assisted CS showed CH_4 conversion above 80% at 700 °C, good stability, and a low coke formation after about 60 h on stream. The catalyst prepared by impregnation showed instead a decrease of CH_4 conversion after 60 h, likely due to deactivation by coke deposition, according to TPO (Temperature Programmed Oxidation) analysis and TEM inspection, which showed the occurrence of carbon nanotubes, not observed when the catalysts were obtained by MW-assisted CS (Figure 5).

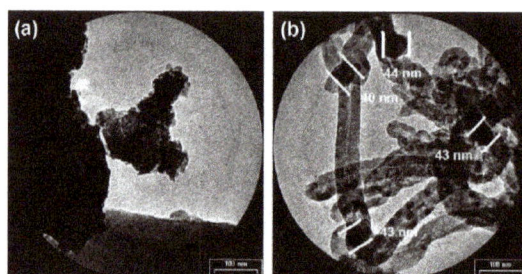

Figure 5. TEM pictures of the catalysts prepared by MW-assisted CS (**a**) and impregnation (**b**) after reaction. Reprinted from *Appl. Catal. A Gen.*, 378, Barros, B. S.; Melo, D. M. A.; Libs, S.; Kiennemann, A., CO_2 Reforming of Methane Over La_2NiO_4/α-Al_2O_3 Prepared by Microwave-Assisted Self-Combustion Method, 69–75. Copyright (2010) with permission from Elsevier.

Moreover, the crystalline structure of the catalysts prepared by MW-assisted CS was preserved after CO_2 reforming of CH_4, whereas a prominent loss of crystallinity occurred in the catalyst obtained by impregnation (which was also deactivated) [51]; the authors ascribed the stability achieved by MW-assisted CS to the strong interaction of Ni active sites with the perovskite structure.

CO_2 selective reduction to light olefins is a process that recently has progressed due to the urgent need of converting CO_2 into building-block chemicals [13,14,52]. To raise selectivity towards light olefins, catalysts that are more effective are needed, since CO_2 is very stable, and FTS usually leads to mixtures of products. K-promoted Fe catalyst is very selective towards CO_2 hydrogenation to light olefins; additionally to K, Zn is also used as a promoter that accelerates FTS, increases catalyst stability, enhances light olefins selectivity, improves CO_2 adsorption, promotes H_2 dissociation, and is highly active in both WGS and RWGS reactions [52]. The effect of MW was studied on the synthesis of HT precipitation followed by incipient wet impregnation (to add K) of a series of K-promoted Fe-Zn catalysts with different Fe/Zn molar ratios. Physico-chemical characterization of the catalysts showed they were characterized by rather uniform particle size (ca. 100 nm). XRD analysis showed the Zn addition to the K-promoted Fe_2O_3 matrix led to formation of $ZnFe_2O_4$ spinel phase and ZnO phase, and also brought about an increase of the surface area. Moreover, the Fe-Zn interaction affected the adsorption and reduction of CO_2 adsorption.

The catalysts exhibited excellent activity for CO_2 hydrogenation and high selectivity toward C_2–C_4 olefins owing to the promoting effect of Zn, due to the fact that particles with uniform shape were obtained, and a fair dispersion of active components occurred with respect to conventional co-precipitation. Additionally, the formed $ZnFe_2O_4$ phase imparted stability to the catalyst, although lowering its activity due to segregation of a fraction of Fe (the active phase).

2.3. MW-Assisted Synthesis of Photocatalysts for CO_2 Reduction

Photocatalytic CO_2 reduction processes (i.e., biomimetic artificial photosynthesis) are very challenging due to intrinsic limitations of the used semiconductors, like poor absorption in the visible range and fast electron/holes recombination rate (e.g., for TiO_2) and stability (e.g., for CdS). "Plasmonic" photocatalysis is a plasmonic metal-enhanced photocatalytic process occurring when noble metal nanoparticles (NPs) decorate the surface of a semiconductor (e.g., TiO_2) to improve its photocatalytic efficiency through the localized surface plasmon resonance (LSPR) effect of NPs [53].

Most of the studies on the subject concern either Au or Ag NPs supported on TiO_2, but their actual use is hampered by the cost and the limited abundance of the noble metals. Recently, it has been shown that also Cu NPs can have LSPR effect, indicating that Cu could replace noble metals.

Cu NPs-decorated TiO_2 nanoflower films (Cu/TiO_2 NFFs) were prepared using both HT method and MW-assisted reduction process. Polyvinylpyrrolidone was used as capping agent and NaH_2PO_2*H_2O as reducing agent in diethylglycole solution under MW irradiation. The catalysts were tested in the photocatalytic reduction of CO_2 to CH_3OH [54].

Physico-chemical characterization of the films showed that both Cu NPs and TiO_2 films were characterized by a fair dispersion of Cu NPs on the TiO_2 NFFs (Figure 6) and by visible light harvesting properties, due to LSPR of Cu NPs and the peculiar nanostructure of the TiO_2 film. Moreover, the presence of Cu NPs allowed fluorescence quenching due to the suppression of charge carriers. The CH_3OH production rate reached 1.8 mmol cm^{-2} h^{-1} (with an energy efficiency of 0.8%) with a 0.5 wt % Cu/TiO_2 NFFs under UV and visible light irradiation: with Cu NPs, the production rate was ca. 6.0 times higher than with pure TiO_2 film (Figure 7).

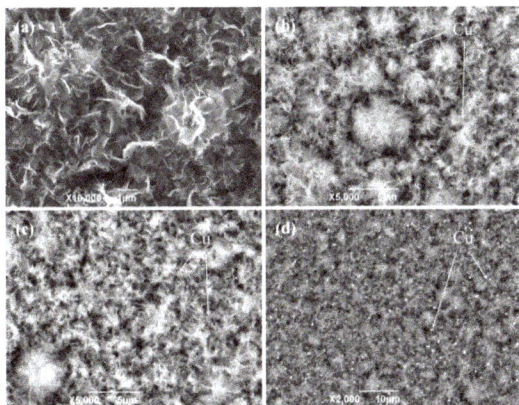

Figure 6. SEM picture of pure TiO_2 film (**a**) and Back Scattered Electrons images of TiO_2 NFFs containing increasing amounts of Cu from (**b**–**d**). Reprinted from *Mater. Res. Bull.*, 68, Liu, E.; Qi, L.; Bian, J.; Chen, Y.; Hu, X.; Fan, J.; Liu, H.; Zhu, C.; Wang, Q., A Facile Strategy to Fabricate Plasmonic, Cu-Modified TiO_2 Nano-flower Films for Photocatalytic Reduction of CO_2 to Methanol, 203–209. Copyright (2015) with permission from Elsevier.

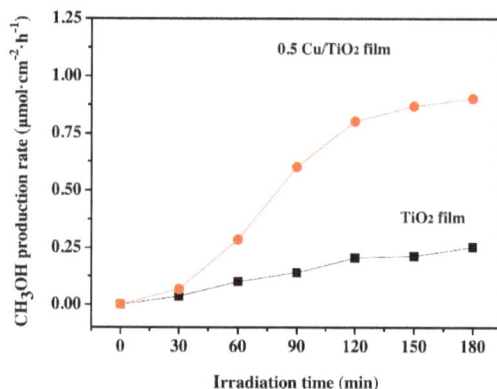

Figure 7. CH$_3$OH production rate by photocatalytic reduction of CO$_2$ as obtained under UV light with both pure TiO$_2$ film and a Cu/NFF obtained by starting from 0.5 mM Cu^{2+} solution (0.5 Cu/TiO$_2$ film). Reprinted from *Mater. Res. Bull.*, 68, Liu, E.; Qi, L.; Bian, J.; Chen, Y.; Hu, X.; Fan, J.; Liu, H.; Zhu, C.; Wang, Q., A Facile Strategy to Fabricate Plasmonic, Cu-Modified TiO$_2$ Nano-Flower Films for Photocatalytic Reduction of CO$_2$ to Methanol, 203–209. Copyright (2015) with permission from Elsevier.

The positive photocatalytic performance towards CO$_2$ reduction to CH$_3$OH observed when Cu NPs were present (Figure 7) was ascribed to a synergistic mechanism of Cu NPs at the surface of TiO$_2$ (Scheme 3): Cu NPs obtained by MW-assisted reduction method likely suppressed the recombination of charge carriers and improved the charge transfer efficiency.

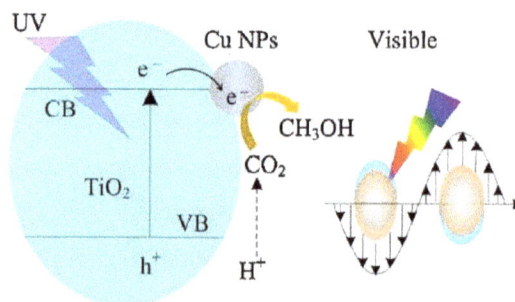

Scheme 3. The synergistic mechanism proposed for the photocatalytic reduction of CO$_2$ to CH$_3$OH with Cu/TiO$_2$ NFFs. Reprinted from *Mater. Res. Bull.*, 68, Liu, E.; Qi, L.; Bian, J.; Chen, Y.; Hu, X.; Fan, J.; Liu, H.; Zhu, C.; Wang, Q., A Facile Strategy to Fabricate Plasmonic, Cu-Modified TiO$_2$ Nano-Flower Films for Photocatalytic Reduction of CO$_2$ to Methanol, 203–209. Copyright (2015) with permission from Elsevier.

In another study, Cu-Graphene Oxide (Cu/GO) composite photocatalysts were obtained using a simple and rapid one-pot MW process [55], shown in Figure 8a.

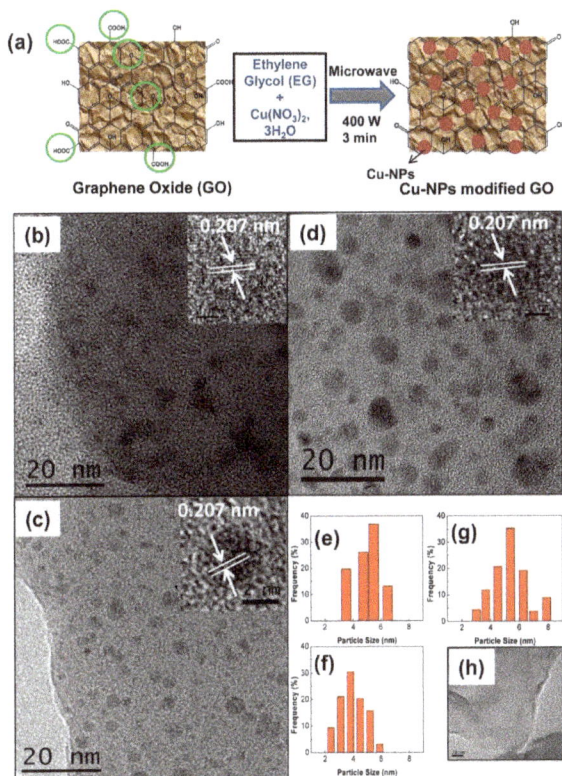

Figure 8. Scheme of the MW assisted synthesis process adopted to decorate GO with Cu NPs (**a**). Sections (**b**–**d**): TEM images of Cu/GO hybrids with 5, 10, and 15 wt % Cu, respectively. Inset: HRTEM image of single Cu-NP of the respective Cu/GO hybrid. Sections (**e**–**g**): size distribution of Cu-NPs in Cu/GO hybrids with 5, 10, and 15 wt % Cu, respectively. Representative TEM image of pure GO (**h**). Reprinted with permission from Shown, I.; Hsu, H.-C.; Chang, Y.-C.; Lin, C.-H.; Roy, P.K.; Ganguly, A.; Wang, C.-H.; Chang, J.-K.; Wu, C.-I.; Chen, L.-C.; Chen, K.-H., Highly Efficient Visible Light Photocatalytic Reduction of CO_2 to Hydrocarbon Fuels by Cu-Nanoparticle Decorated Graphene Oxide. *Nano Lett.* 14, 6097–6103. Copyright (2014) American Chemical Society.

By using $Cu(NO_3)_2$ as metal precursor, short MW heating (ca. 3 min at 400 W) was performed in ethylene glycol solution (Figure 8a), leading to a fair dispersion of crystalline Cu NPs in the GO.

The obtained Cu NPs (with a size of ca. 4–5 nm, Figure 8) were strongly interacting with GO: the obtained composite was particularly active under solar irradiation in the production of both CH_3OH and acetaldehyde. Cu NPs were shown to enhance the GO photocatalytic activity of GO, essentially by suppressing electron-hole recombination, and also by reducing the GO bandgap and by modifying its work function (Scheme 4). The presence of Cu NPs greatly enhanced absorption of GO in the visible range, though the shift in the band gap by effect of Cu NPs was not well understood. XPS analysis also showed that oxygenated groups of GO (i.e., -COOH and C-O-C groups) were almost completely removed by the MW treatment. No other Cu oxides were detected, indicating that the MW heating led to the reduction of all the Cu ions.

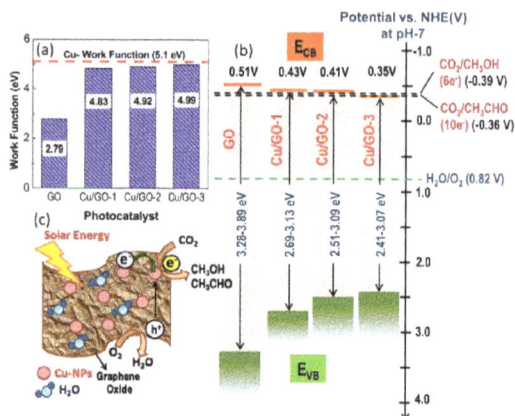

Scheme 4. Section (**a**): work functions of pure GO and of Cu/GO hybrids with 5, 10, and 15 wt % Cu (Cu/GO-1, Cu/GO-2, and Cu/GO-3, respectively). Section (**b**): band-edge positions of pure GO and Cu/GO hybrids in comparison with CO_2/CH_3OH and CO_2/CH_3CHO formation potentials. Section (**c**): proposed photocatalytic reaction mechanism under solar irradiation. Reprinted with permission from Shown, I.; Hsu, H.-C.; Chang, Y.-C.; Lin, C.-H.; Roy, P.K.; Ganguly, A.; Wang, C.-H.; Chang, J.-K.; Wu, C.-I.; Chen, L.-C.; Chen, K.-H. Highly Efficient Visible Light Photocatalytic Reduction of CO_2 to Hydrocarbon Fuels by Cu-Nanoparticle Decorated Graphene Oxide. *Nano Lett.* **14**, 6097–6103. Copyright (2014) American Chemical Society.

XPS analysis [55] showed the likely occurrence of a charge transfer from GO to Cu NPs, and a correlation was found between the Cu content and the rate of formation and selectivity of the products. The Cu/GO catalyst with 10 wt % Cu was more than 60 times more active than pure GO and ca. 240 times more active than commercial TiO_2 (Degussa P25) photocatalyst under visible light.

Also, Cu_2O can be used for CO_2 photoconversion: An et al. [56] produced Cu_2O/RGO (Reduced Graphene Oxide) composites using a facile, one-step, MW-assisted HT synthesis. X-ray diffraction showed that the Cu_2O was likely the only Cu-containing phase in the composites, and MW-assisted synthesis allowed the formation of composites where intimate contact between RGO sheets and Cu_2O microspheres (a result of the strong affinity between the metal oxide and the abundant functional groups of GO) was enhanced. Such strong interaction between Cu_2O and RGO was attributed to the adopted MW-assisted synthesis: such interaction was very effective in that RGO positively affected the photocatalytic activity, improving both activity and stability of Cu_2O, due to an efficient charge separation and transfer to RGO (Scheme 5).

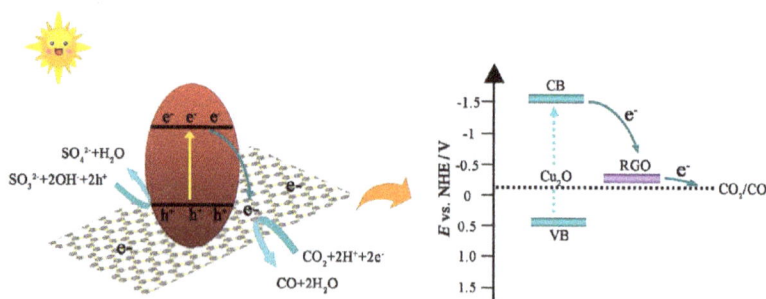

Scheme 5. Schematic picture of the charge transfer process in Cu_2O/RGO composites during CO_2 photoreduction to CO in water. Reprinted from Ref. [56]. Copyright 2014 with permission of Wiley.

Cu-decorated graphitic carbon nitride (g-C_3N_4) nanosheets were obtained by a procedure depicted in Scheme 6, in which the final step involved the MW-assisted HT synthesis of Cu NPs [57].

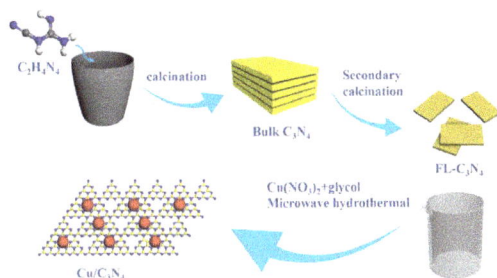

Scheme 6. Schematic picture of the synthesis of Cu decorated g-C_3N_4 nanosheets. Reprinted from *Appl. Surf. Sci.*, 427, Shi, G.; Yang, L.; Liu, Z.; Chen, X.; Zhou, J.; Yu, Y., Photocatalytic Reduction of CO_2 to CO Over Copper-Decorated g-C_3N_4 Nanosheets with Enhanced Yield and Selectivity, 1165–1173. Copyright (2014) with permission from Elsevier.

The composites were tested for the photocatalytic reduction of CO_2 to CO. During MW-assisted HT synthesis, crystalline Cu NPs were obtained: some of them were deposited at the outer surface of nanosheets, and other were wrapped within them, as shown by TEM pictures in Figure 9. Such an intimate contact between the two phases likely enhanced the transfer of electrons and positively affected the photocatalytic activity.

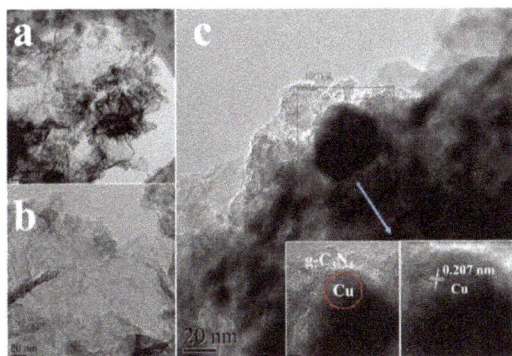

Figure 9. TEM images of g-C_3N_4 at low (**a**) and high (**b**) magnification and TEM image of Cu decorated g-C_3N_4 nanosheets (**c**). Reprinted from *Appl. Surf. Sci.*, 427, Shi, G.; Yang, L.; Liu, Z.; Chen, X.; Zhou, J.; Yu, Y., Photocatalytic Reduction of CO_2 to CO Over Copper-Decorated g-C_3N_4 Nanosheets with Enhanced Yield and Selectivity, 1165–1173. Copyright (2014) with permission from Elsevier.

The presence of Cu NPs also affected the ability of absorbing visible light, as shown by DR UV-Vis spectroscopy: Cu NPs likely facilitate the light harvesting process; moreover, the absorption edge of the composites shifted towards longer wavelength, indicating that a higher number of electrons/holes could be generated.

Photocatalytic tests showed that Cu had a positive effect of the CO yield. An optimal amount of Cu was found, corresponding to 6 wt % Cu. That content was a kind of tradeoff between 3 wt % (a too low Cu content) and 9 wt %, in which a higher number of Cu NPs can became centers for electrons/holes recombination. Scheme 7 reports the proposed mechanism, in which photogenerated electrons excited in g-C_3N_4 migrate to the surface of Cu NPs, acting both as trapping centers and catalytic sites, for the photocatalytic CO_2 reduction to CO. Simultaneously, photogenerated holes migrated to the surface of the photocatalyst for water oxidation to O_2.

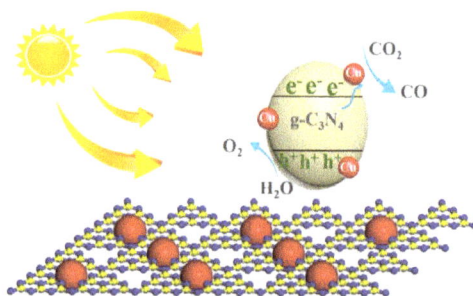

Scheme 7. Scheme of the photocatalytic reduction of CO_2 to CO and simultaneous H_2O oxidation to O_2 in the presence of Cu-decorated g-C_3N_4 nanosheets. Reprinted from *Appl. Surf. Sci.*, 427, Shi, G.; Yang, L.; Liu, Z.; Chen, X.; Zhou, J.; Yu, Y., Photocatalytic Reduction of CO_2 to CO Over Copper-Decorated g-C_3N_4 Nanosheets with Enhanced Yield and Selectivity, 1165–1173. Copyright (2014) with permission from Elsevier.

CdS is one of the most promising candidate semiconductors as photocatalysts active in the Vis range: Yu et al. [58] prepared RGO–CdS nanorods composites in ethanolamine–water solution using a one-step MW-assisted HT method. The composites exhibited a high activity in the photocatalytic reduction of CO_2 to CH_4, even without noble metal catalyst (i.e., Pt, which is usually adopted as co-catalyst). MW-assisted HT synthesis requires shorter time and less energy than conventional HT synthesis. Furthermore, particles with uniform size are usually obtained. Raman and XPS spectroscopies confirmed the reduction of parent GO using the MW-assisted method. TEM analysis (Figure 10) showed an intimate connection between the two materials.

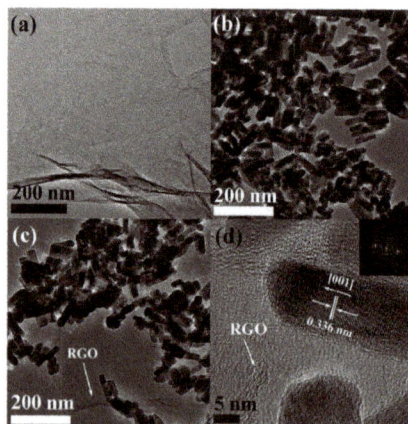

Figure 10. TEM images of pure GO (**a**) and pure CdS nanorods (**b**). TEM (**c**) and HRTEM (**d**) images of RGO-CdS nanorods composites. Reproduced from Ref. [58]. Copyright 2014 with permission from The Royal Society of Chemistry.

An optimal RGO content of 0.5 wt % was found, leading to the highest CH_4 yield. Higher RGO content led to lower CH_4 yields; the authors ascribed this result to RGO shielding light absorption by CdS nanorods. RGO acted as co-catalyst, enhancing the photocatalytic reduction of CO_2 to CH_4, with the best performance attained with a RGO content of 0.5 wt %; the corresponding CH_4 production rate was 2.51 mmol h^{-1} g^{-1}, which exceeded that of pure CdS nanorods by more than 10 times. According to the authors, RGO was able to adsorb more CO_2 molecules and to activate them very efficiently, simultaneously generating a higher number of photogenerated electrons on its surface, finally leading to an enhancement of CO_2 reduction performance of the RGO–CdS composite (Scheme 8).

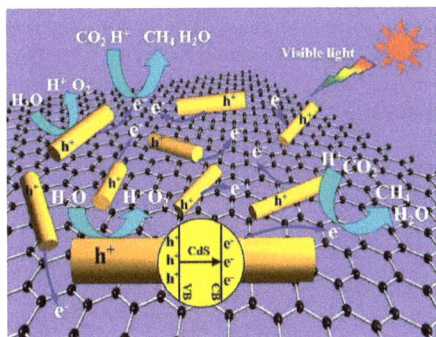

Scheme 8. Scheme of the mechanism proposed for the photocatalytic reduction of CO_2 to CH_4 by RGO-CdS nanorods composites. Reprinted from Ref. [58]. Copyright 2014 with permission from The Royal Society of Chemistry.

In another work, MW were found to positively affect the immobilization of homogeneous Co(III) cyclam complex (cyclam is 1,4,8,11-tetraazacyclotetradecane, namely: [Co(cyclam)Cl_2]Cl) at the surface of SBA-15 silica. The photocatalytic activity towards CO_2 reduction to CO was studied in the presence of *p*-terphenyl as a photosensitizer [59]. Usually, Co(III) cyclam immobilization is performed by reflux method, but MW-assisted immobilization of the Co(III) cyclam complex occurred more rapidly and led to more uniform Co(III) surface species. Moreover, a smaller amount of Co(III) complex was needed with respect to reflux method. This latter aspect led to well-defined, molecularly dispersed Co(III) sites within SBA-15 mesopores. IR spectroscopy showed that the complex interacts with the silica surface mainly by chemisorption on silanol ($\equiv$SiOH) groups. The photocatalyst prepared by MW-assisted synthesis was more active and selective than that prepared by reflux method, in the presence of a molecular photosensitizer and sacrificial electron donors (Figure 11).

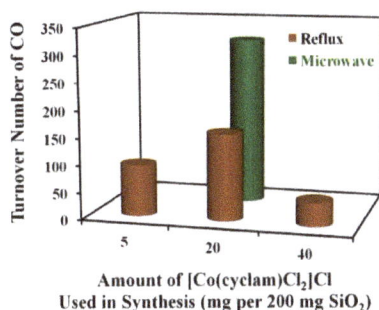

Figure 11. CO production by CO_2 photochemical reduction after 4 h reaction with 2.0 mM p-terphenyl with Co(III) cyclam deposited by MW-assisted synthesis and conventional reflux method. Reprinted from Ref. [59]. Copyright 2017 with permission from The Royal Society of Chemistry.

3. Ultrasound as Unconventional Activation Method

3.1. General Principles of US-Assisted Synthesis of (Nano)Materials

US is essentially a mechanical longitudinal wave with frequency equal to or higher than 20 kHz, which is the upper limit of human hearing range. US propagation gives sinusoidal variation in the static pressure in the liquid medium inducing cavitation, a phenomenon controlled by external parameters, i.e., US frequency and intensity and type solvent. Cavitation is based on radical and mechanical effects. The radical effect occurs either at the bubble interface or in the interior cavity and is ascribed to the occurrence of the sonolysis of molecules. The mechanical effect, occurring after the cavity collapse, arises from shear forces, microjets, and shock waves outside the bubble, therefore producing severe physical changes in the presence of solids or metals. The transient implosive collapse of the bubble is an adiabatic phenomenon, during which inside the bubble very high temperature and pressure occur (~4600 °C and ~50 MPa). The phenomenon affects the reaction system [60] by generating both an intense local convection in the reaction medium (finally enhancing the mass transfer processes) and highly reactive radical species within the bubbles, due to high local temperature and pressure. Upon fragmentation of the bubble, the radicals generated therein get released into the medium, where they can induce/accelerate chemical reactions [61].

On the one hand, US has emerged as a potential green technique for the intensification of different physical, chemical, and biological processes. Compared with conventional heating, the use of US irradiation reduced the reaction time from hours to minutes. On the other hand, US can be effectively employed in the synthesis of NPs with a high production yield, without using expensive instruments and/or extreme conditions [62,63], as, for example, carbon encapsulated Fe NPs with superior catalytic activity and reusability for hydrogenation reaction in the liquid phase [64]. US allow obtaining of highly reactive surfaces, enhancing mixing and increasing mass transport. Therefore, such unconventional method can be particularly helpful for preparing and activating catalytic materials [65].

3.2. US-Assisted Synthesis of Cu–ZnO Catalysts and of Their Improved Formulations

Under typical industrial conditions, CH_3OH is formed mostly via CO_2 hydrogenation, with CO acting as CO_2 source and as oxygen scavenger from water, which inhibits the active metal sites [66]. A synergism between Cu^0 and Cu^+ sites is the most adopted explanation for the activity of Cu/ZnO-promoted catalysts [38,61,67,68]; however, Cu^0 sites [69,70] and/or Cu–Zn alloy [71–73] have been proposed to have a catalytic role as well.

Many unconventional preparation routes have been adopted to obtain Cu–ZnO catalysts with improved activity, selectivity, resistance to poisons, and enhanced lifetime [38,74–78]. In this frame, the SG method was largely adopted in order to increase the metal surface area along with the total surface area [79,80]. Moreover, the excellent activity of oxalate co-precipitated catalysts was also demonstrated [81], and Cu–Zn catalysts prepared by "spark-erosion" showed superior performance in terms of selectivity [82], with noticeable TOF in the CO_2 hydrogenation reaction that was ascribed to a close contact among the phases in $Cu/ZnO/ZrO_2$ catalyst produced by combustion.

To further improve the physico-chemical properties of a $Cu–ZnO/ZrO_2$ catalyst (Zn_{at}/Cu_{at}, 0–3; ZrO_2, 42–44 wt %), a new synthetic procedure consisting in the US-assisted reverse co-precipitation was proposed [83]. The US irradiation guaranteed highly increased total surface area (BET, 120–180 m^2 g^{-1}), although the Zn/Cu atomic ratio strongly affected either the specific surface area or the volume of the pores. Indeed, the materials with low ($\leq$0.1) and high (>0.7) Zn/Cu ratios have lower surface area than those with a Zn/Cu ratio of 0.3–0.4, corresponding to a maximum of 175 m^2/g. An analogous volcano trend was observed for the pore volume and pore diameter, in which the catalyst with Zn/Cu ratio equal to 0.4 contains mainly mesopores with diameter in the 70–120 Å range and, as a consequence, an enhanced cumulative volume (0.6 cm^3 g^{-1}). Moreover, the authors found that the US irradiation led to very high dispersion (3–58%) and exposure of Cu (S_{Cu} = 9–63 m^2 g^{-1}). The catalytic activity in the CH_3OH production by CO_2 hydrogenation was explored in a range of temperatures (160–260 °C) and pressures (1.0–3.0 MPa), and compared with a commercial Cu–ZnO/Al$_2$O$_3$ catalyst with optimum results ascribed to the presence of highly dispersed Cu particles.

The direct synthesis of dimethyl ether starting from syngas involves two consecutive reactions, i.e., CH$_3$OH synthesis and CH$_3$OH dehydration [84]. Alternatively, dimethyl ether can be directly obtained from syngas or CO$_2$ hydrogenation.

The use of novel bi-functional catalysts based on metal oxides (namely, CuO, ZnO, Al$_2$O$_3$, and possibly some precious metal oxides, such as Cr$_2$O$_3$) to synthesize CH$_3$OH and on acid solids (namely, HZSM-5 and HY zeolites or SAPOs) to convert it into dimethyl ether can significantly enhance the yield with respect to the two-step process, because CH$_3$OH dehydration can interfere with the equilibrium of the synthesis reaction [85].

In this contest, Cu-ZnO-Al$_2$O$_3$, as well as Al$_2$O$_3$ and HZSM-5 catalysts, have been widely investigated [86–88]. In particular, a previous study reported on the preparation of CuO-ZnO-Al$_2$O$_3$/HZSM-5 hybrid catalysts (CZA/HZSM-5) using different techniques, such as impregnation, co-precipitation-physical mixing method, and a US-assisted co-precipitation procedure [89]. It was observed that the CZA/HZSM-5 catalysts prepared by US-assisted co-precipitation displayed higher activity than the systems obtained by conventional physical mixing and impregnation. Further, various catalysts with different CZA and HZSM-5 wt. ratio (namely, 2, 3, and 4) were synthesized by the US-assisted co-precipitation procedure summarized in Figure 12 in order to establish a correlation between the catalyst composition and the displayed activity [90]. The experimental conditions for US irradiation in Ar atmosphere were 90 W for 45 min, with 10 min current pulse time and 0.5 min rest time, by means of the experimental apparatus shown in Figure 13.

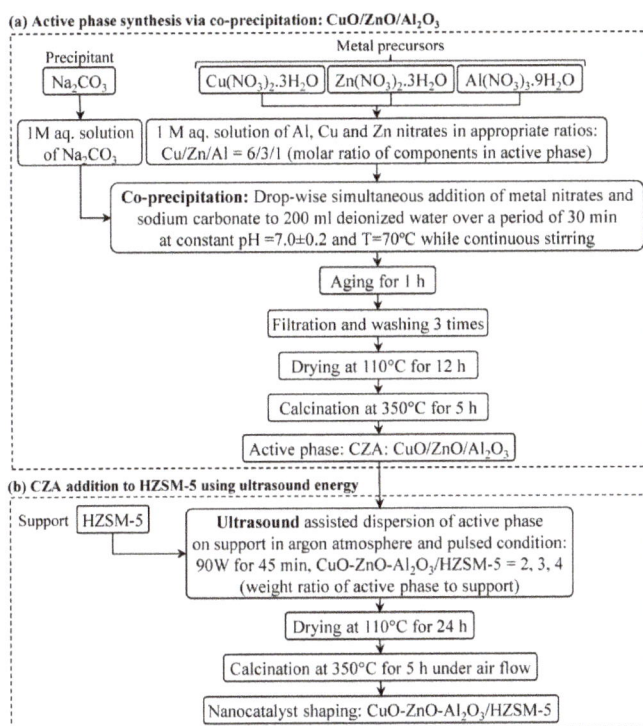

Figure 12. Schematic flow chart for preparation steps of CZA/HZSM-5 catalysts synthesized via combined co-precipitation-ultrasound method with different active phase to support ratios. Reprinted from Ref. [90]. Copyright 2014 with permission of The Royal Society of Chemistry.

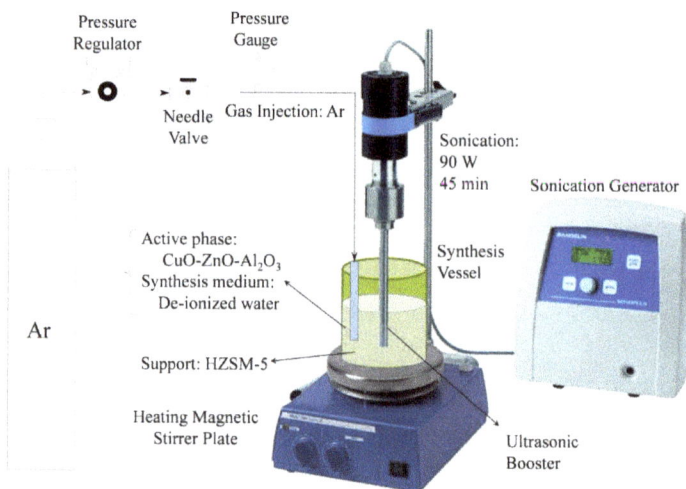

Figure 13. Experimental setup for dispersion of CZA over HZSM-5 using US energy. Reprinted from Ref. [90]. Copyright 2014 with permission of The Royal Society of Chemistry.

It is worth noting that the zeolite preserved its crystal structure during the preparation. Moreover, the crystallinity was significantly influenced by varying of the CZA content. XRD peaks related to monoclinic CuO, and hexagonal ZnO phases were observed. FESEM analyses indicated the presence of a uniform coating of the HZSM-5 surface by CZA upon US irradiation during the synthesis. In addition, the catalysts consisted of many small particles with almost spherical shape. At variance with reducibility, the surface area was not dependent on the CZA content. The HZSM-5 structure was not damaged upon loading CZA, as demonstrated by FTIR spectroscopy results. The optimal composition of the CZA/HZSM-5 catalysts for the conversion of syngas to dimethylether was 4:1 at 275 °C and 40 bar.

Interestingly, it was reported that the sonication time significantly affected both morphology and structure of the CZA/HZSM-5 catalyst [91]. In particular, larger CuO crystals formed at longer irradiation times, and at the same time the particle aggregates changed into smaller size and more uniform shape, resulting in an increase of the surface area. The catalyst synthesized under long US irradiation showed very good activity during time on stream test, whilst the non-sonicated sample lost its activity.

The effect of US irradiation power was evaluated on different CZA/HZSM-5 catalysts prepared by US-assisted co-precipitation method [92]. The results of a thorough physico-chemical characterization evidenced that the nature of the precursors and the time of irradiation strongly affected the morphology, the surface area, the functional groups, and the structure of the final materials. In particular, as shown in Figure 14, highly irradiated acetate-based catalysts had uniform morphology, as well as smaller CuO particles with higher surface area, due to high nucleation rate and high cooling rates, resulting in a stronger interaction among the components in comparison to the catalysts prepared by starting from nitrate precursors.

It is worth noting that an increase of the irradiation power gave rise to smaller particles, resulting in an improved dispersion and surface area of the catalysts. Moreover, CuO particles with lower crystallinity were generated at high power, resulting in an enhanced interaction among the metal oxide particles. In addition, such ultrasound-assisted co-precipitation method allowed to produce CZA/HZSM-5 catalysts that were more stable with respect to those prepared by conventional precipitation, for which 18% CO conversion and 58% dimethyl ether yield decreases in 24 h time on stream test were observed.

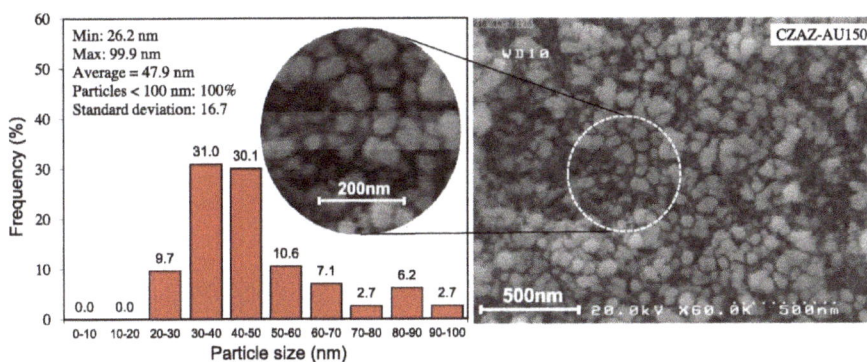

Figure 14. Surface particle size distribution histogram of synthesized CZA/HZSM-5 nanocatalyst using acetate precursor and ultrasound irradiation power of 150 W (CZAZ-AU150). Reprinted from *Ultrason. Sonochem.*, 21, Allahyari, S.; Haghighi, M.; Ebadi, A.; Hosseinzadeh, S., Ultrasound-Assisted Co-Precipitation of Nanostructured CuO-ZnO-Al_2O_3 Over HZSM-5: Effect of Precursor and Irradiation Power on Nanocatalyst Properties and Catalytic Performance for Direct Syngas to DME, 663–673. Copyright (2014) with permission from Elsevier.

The authors proposed a pathway for the ZnO synthesis under ultrasound irradiation in CZA that is illustrated in Figure 15. Such pathway can be extended to other precursors, as well as to CuO and Al_2O_3. More specifically, the metallic precursors underwent decomposition and hydratation upon ultrasound irradiation, and H· and OH· radicals formed by water sonolysis [93]. The free OH· radicals formed in the cavities or at interface can react with the metal cations forming metal hydroxides that are converted to the corresponding metal oxides upon thermal treatment. Simultaneously, by co-precipitation of the metal precursor, the metal carbonates (that will then be converted into the metal oxides) are produced. Therefore, a strong synergistic action of US during co-precipitation was advised. In particular, the OH· radicals coming from the water sonolysis are able to convert a fraction of the metal salt to the corresponding metal oxide without any help from the precipitant. Moreover, a significant increase of the nucleation rate under peculiar temperature and pressure conditions upon cavity collapse gives rise to the formation of small and well-dispersed NPs. In particular, the irradiation cycles impede the growth of the nucleated particles, resulting in the production of small catalyst particles. Finally, particle agglomeration is prevented, indeed, the populations of OH· radicals and of the nucleation sites are enhanced at high US power, and the conversion to metal oxides is therefore improved.

An US-assisted co-precipitation method was employed for the preparation of CZA/HZSM-5 catalysts supported on multi-walled carbon nanotubes pretreated with H_2SO_4/HNO_3 mixture [94], in which the acid pretreatment had the aim of activating and functionalizing the surface of the nanotubes. The acid-pretreated carbon nanotubes were dispersed in deionized water by sonication more easily than the pristine ones, because the presence of the surface functional groups increased the capability to form hydrogen bonds, converting the surface of the multi-walled carbon nanotubes from hydrophobic to hydrophilic, therefore resulting in improved dispersion and suspension stability. The CZA NPs of about 10–20 nm were quite homogeneously distributed and densely deposited at the surface of the acid-pretreated, multi-walled carbon nanotubes (as shown in Figure 16).

The catalysts were tested in the hydrogenation of CO_2 to produce dimethyl ether in a fixed-bed reactor. It was shown that the presence of the multi-walled carbon nanotubes has a promoting effect on the catalytic activity and on the dimethyl ether. The CO_2 conversion was 46.2%, with dimethyl ether yield and selectivity of 20.9% and 45.2%, respectively, at 262 °C, 3.0 MPa, volume ratio H_2/CO_2 = 3, and space velocity = 1800 mL $g_{cat}^{-1}h^{-1}$.

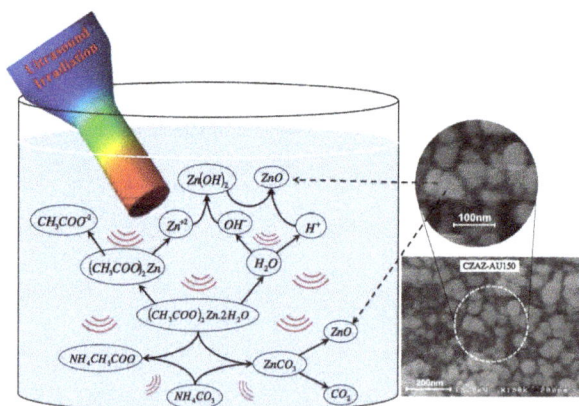

Figure 15. Reaction pathway for synthesis of ZnO in CZA nanocatalyst by ultrasound-assisted co-precipitation of zinc acetate. Reprinted from *Ultrason. Sonochem.*, 21, Allahyari, S.; Haghighi, M.; Ebadi, A.; Hosseinzadeh, S., Ultrasound assisted co-precipitation of nanostructured CuO-ZnO-Al$_2$O$_3$ over HZSM-5: Effect of Precursor and Irradiation Power on Nanocatalyst Properties and Catalytic Performance for Direct Syngas to Dimethyl Ether, 663–673. Copyright (2014) with permission from Elsevier.

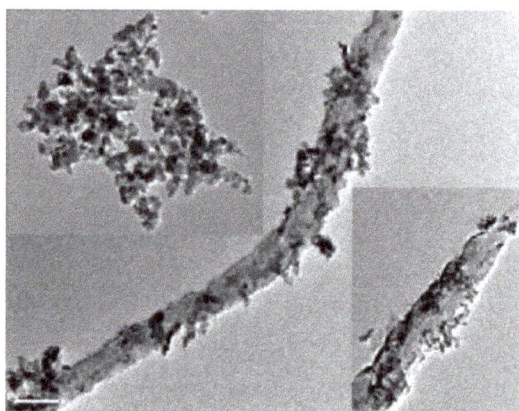

Figure 16. TEM images of CZA supported on multi-walled carbon nanotubes pretreated with a H$_2$SO$_4$/HNO$_3$ mixture. Reprinted from *Appl. Surf. Sci.*, 285, Zha, F.; Tian, H.; Yan, J.; Chang, Y., Multi-Walled Carbon Nanotubes as Catalyst Promoter for Dimethyl Ether Synthesis from CO$_2$ Hydrogenation, 945–951. Copyright (2013) with permission from Elsevier.

3.3. US-Assisted Synthesis of Catalysts with Ferromagnetic Properties

The synthesis of nanostructured materials based on ferromagnetic elements is of great interest due to the possibility to tune their magnetic properties [95–97], as well as to promote physico-chemical surface processes [98,99] by controlling the shape and size of the obtained NPs. In addition, magnetic catalysts can be easily positioned and recovered by using an external magnetic field.

In a recent paper, Vargas et al. reported a facile two-step UP procedure to synthesize Ni NPs and then support them on *m*-ZrO$_2$ to obtain low-cost and active catalysts for CO$_2$ methanation (Equation (2)) [100]. More in detail, Ni NPs were firstly produced by chemical reduction of a metal precursor (nickel chloride) with either hydrazine hydrate (N$_2$H$_4$·H$_2$O) or sodium borohydride (NaBH$_4$)

in water, in the presence of citric acid and sodium citrate as protecting agents. The chemical reduction was carried out under US irradiation (40 kHz, 300 W) without any additional stirring sources. Then, the black colored solution, indicating the presence of Ni^0 NPs, was added to other aqueous suspensions of commercial monoclinic ZrO_2. In order to obtain homogenous materials, the obtained mixtures were maintained overnight in a conventional ultrasonic bath at ambient conditions. The two samples were centrifuged, washed several times with a mixture of deionized water and ethanol, and dried under vacuum. The catalysts had the same metal loading (1 wt % Ni), but different metal particle sizes. It was found that the final particle size was strongly related to the employed reducing agent. Particularly, homogeneous Ni particles with globular shape and average size of 40 ± 1 nm formed in the presence of hydrazine, as shown in Figure 17A.

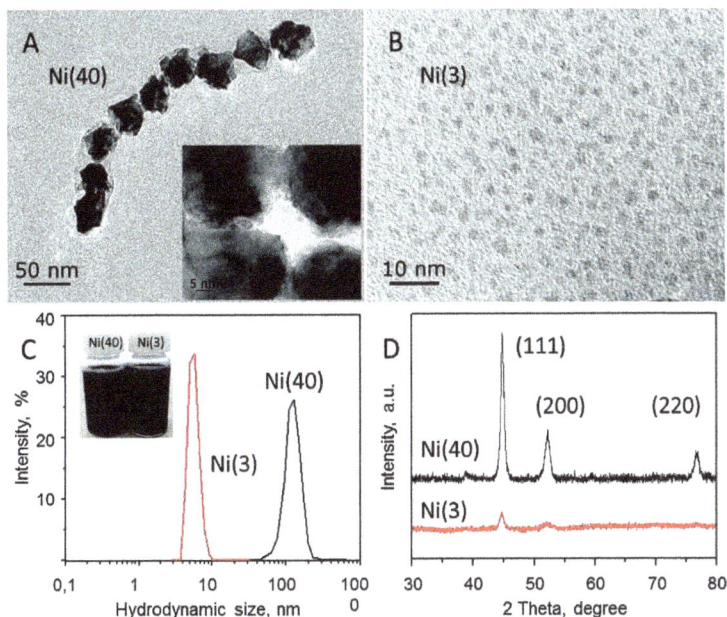

Figure 17. TEM images of nickel nanoparticles developed using hydrazine (**A**), high magnification image (inset), and sodium borohydride (**B**). Hydrodynamic diameter (**C**), colloidal systems images (inset), and XRD patterns (**D**) of samples. Reproduced from [100] with permission of The Royal Society of Chemistry.

On the contrary, when $NaBH_4$ was employed, smaller and well isolated Ni NPs with average diameter equal to 2.7 ± 0.4 nm were observed (Figure 17B). Such differences can be explained by considering that in the case of N_2H_4*H_2O, the formation of the $[Ni(N_2H_4)_3]^{2+}$ complex implied a slower kinetics, hence inhibiting the formation of nuclei and favoring particle growth. Instead, the use of sodium borohydride gave rise to a larger number of Ni nuclei and, consequently, to smaller particles. Hydrodynamic diameters larger than those measured by TEM were detected in both samples (Figure 17C) in agreement with the presence of citrate at the surface of NPs and of the occurrence of dynamics dipolar magnetic interactions. Peaks at 44.7°, 52.1°, and 76.7° 2 Theta degrees were observed in the diffraction pattern of the catalyst with large Ni particles (Ni(40) in Figure 18c), assigned to the (111), (200), and (220) reflections of the cubic Ni phase (Figure 17D). As for the catalyst with small NPs, (Ni(3) in Figure 18c), only the (111) peak was detected. The catalysts were tested in the CO_2 methanation reaction, revealing high activity (70%) and excellent selectivity to CH_4 (90%) at 720 K.

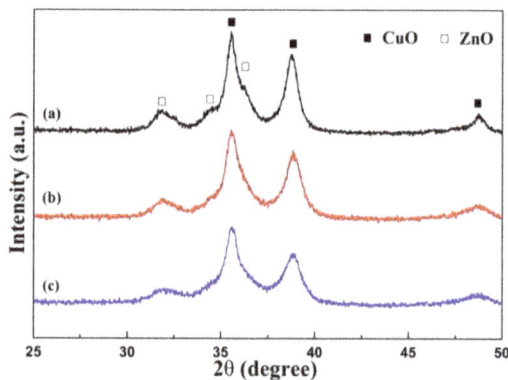

Figure 18. XRD patterns of CZAx catalysts (x = 200 (**a**), 300 (**b**), and 400 (**c**)). Reprinted from Wu, W.; Xie, K.; Sun, D.; Li, X.; Fang, F. CuO/ZnO/Al₂O₃ Catalyst Prepared by Mechanical-Force-Driven Solid-State Ion Exchange and Its Excellent Catalytic Activity under Internal Cooling Condition. *Ind. Eng. Chem. Res.* 56, 8216–8223. Copyright (2017) American Chemical Society.

Ni/ZrO_2 catalysts modified with rare earth elements, in particular Ni/(Zr-Sm)Ox catalysts, showed improved activity in the CO_2 methanation (Equation (2)) [101,102]. Conventional wet impregnation method or preparation starting from amorphous Ni-Sm alloys were the preparation procedures usually adopted for these catalysts. In particular, the activity of these systems was correlated with the presence of (Zr–Sm)Ox units with the tetragonal ZrO_2 structure. Moreover, the catalytic performance was also influenced by preparation method, surface area, and pore size distribution of the catalysts [101,103].

A Ni-based (40 at % Ni) catalyst supported on mesoporous yttria-stabilized-zirconia composite was prepared by a US method using sodium dodecyl sulfate as a templating agent to obtain high surface area (193 m^2/g) [104]. The application of the US method resulted in short reaction time (6 h) and no requirement for the glycolation of nickel, yttrium, and zirconium ions. In another paper by the same research group, a simple one-step US-assisted synthesis (in which the corresponding inorganic salts were used as precursors) was applied to prepare 20–40 at % Ni supported on mesoporous ZrO_2 [105]. In these catalysts, Ce and Sm rare earth ions were introduced in the support to further enhance its activity towards CO_2 methanation. The obtained catalysts possessed in situ generated mesopores, and the Ni oxide (or hydroxide) that were about 10 nm were incorporated into tetragonal ZrO_2 modified by the rare earth metals. In particular, the catalyst containing 30 at % Ni loading had maximum porous volume and size, and at the same time displayed the highest activity, which was further enhanced upon proper oxidation–reduction activation treatment, during which additional active centers were produced.

CO_2 methanation was also performed over Ni/Al_2O_3 catalysts prepared by a US-assisted co-precipitation [106]. The authors observed that an increase in the Ni loading from 5 wt % up to 25 wt % produced a surface area enhancement, resulting in better catalytic performances, and simultaneously decreased the crystallinity and improved the catalyst reducibility. A further increase of the metal loading had a negative effect on the surface area, as well as on the catalytic activity, due to the drop of the Ni dispersion. The 25 wt % Ni/Al_2O_3 catalyst attained high CO_2 conversion (74%) and almost complete CH_4 selectivity (99%) at 350 °C. The catalyst demonstrated also high stability at the same temperature after 10 h time-on-stream. The catalytic performance was observed to decrease at higher Gas Hourly Space Velocity, due to both reduced contact time between the reactants and the catalyst surface, as well as the amount of adsorbed reactants. Conversely, a positive effect on the catalyst activity was observed when higher H_2/CO molar ratio was employed. Furthermore, the performance of catalysts calcined at higher temperature was observed to decrease.

In another recent study, Co/SBA-15 and Co/SBA-16 catalysts with 20 wt % Co loading were prepared by conventional incipient wetness impregnation method [107]. Additional vacuum and US treatments were performed after the synthesis. N_2 physisorption, small angle XRD (SAXRD), wide angle XRD, SEM and energy dispersive X-ray spectroscopy (SEM/EDX), TEM, and TPR were employed to characterize the catalysts. It was observed that Co dispersion on the SBA-15 support was enhanced upon such post-synthesis treatments, due to the decrease of the metal crystallite size. The presence of Co_3O_4 and of different various Co_xO_y species was detected. Both post-impregnation treatments affected the metal-support interaction, increasing the activity in the CO_2 hydrogenation. It was observed that the vacuum treatment gave rise to a material in which crystalline cobalt oxide was located in the inner pores, whilst cobalt oxide agglomeration at the outer surface took place upon the US treatment.

A series of 0.5–5.0% Ru catalysts supported on Al_2O_3 for the selective methanation of CO in H_2-containing streams were investigated with the aim to clarify the influence of process parameters, such as the metal loading, the temperature of calcination, the chlorine ions content, and the space velocity, on the catalytic performances [108]. Careful characterization by XRD, Thermal Gravimetry/Differential Thermal Gravimetry (TG/DTG), and SEM was performed. A number of laboratory experiments were performed to clarify the influence of process parameters (Ru loading, calcinations temperature, space velocity, and chlorine ions content) on the activities of catalysts. It was observed that by carrying out the synthesis of the catalysts by US-assisted impregnation, the catalytic activity was significantly improved, and the impregnation time was significantly reduced. In particular, agglomeration accompanied by non-homogeneous distribution of Ru deposits on Al_2O_3 were observed when the catalysts were prepared by wetness impregnation method. Conversely, US impregnation strongly decreased the particle size and improved the distribution of Ru during the impregnation. The best results were achieved by the catalyst containing 2 wt % Ru on Al_2O_3 calcined at 400 °C. Long-term tests demonstrated high activity and stability under realistic reaction conditions.

4. The Role of Mechanochemical Activation in the Synthesis of Nanostructured Catalysts

4.1. General Principles of MC

MC induces chemical transformations by mechanical action (compression, shear or friction) that allows breaking intramolecular bonds and further chemical reactions. More in detail, radical species are mechanochemically produced due to the breaking of weak bonds and to the extreme surface plasma conditions generated by the mechanical impact. Mechanochemical activation processes have a long history, and their importance does not decrease with time, because many solid-phase reactions (with stoichiometric amount of reactants) can be quantitatively and rapidly promoted under solvent-free conditions [109]. Therefore, the mechanochemical mixing process can be an environmentally advantageous alternative to conventional chemical syntheses, in which solvents often play a key role in energy dispersion and chemicals solvation and transportation [110].

The preparation of nanomaterials for advanced applications by mechanochemical synthesis is among the most suitable alternative routes because of its simplicity and high reproducibility. Indeed, such methodology is also environmentally advantageous in comparison to conventional preparation procedures, since it is often solvent-free, and it is carried out under mild reaction conditions and involving short reaction times.

4.2. MC as a Tool for the Synthesis of Catalysts

The MC synthesis of heterogeneous catalysts strongly affects their catalytic properties. Indeed, the excess of potential energy produced during grinding, together with shear and friction forces, can deeply modify the materials by producing a large variety of defects, hence improving or modifying the reactivity [109,111].

A variety of heterogeneous catalysts, ranging from supported NPs to nanocomposites has been obtained by using MC synthesis, as summarised in Scheme 9 [109].

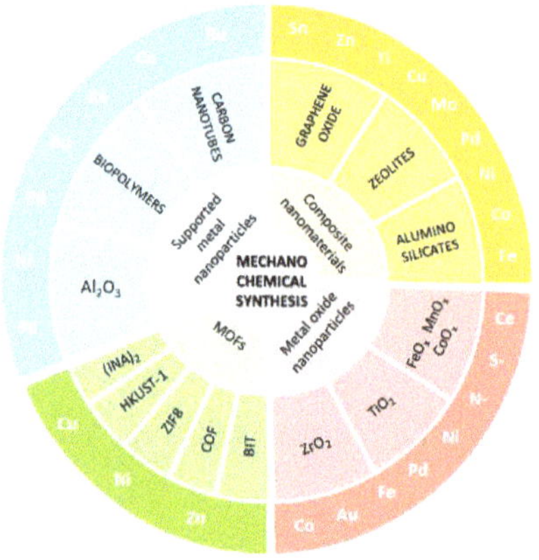

Scheme 9. Pictorial representation of different types of MC synthesized (nano)materials. From supported metal NPs, composite nanomaterials to metal oxide NPs and metal organic frameworks (MOFs) including covalent organic frameworks (COF). Examples of relevant materials include zinc-based ZIF-8 structures (BIT-11), Cu(INA)$_2$, and Cu-containing HKUST-1, as well as supported noble metals on aluminosilicates, graphene, etc. Reprinted from Ref. [109]. Copyright 2015 with permission of the Royal Society of Chemistry.

MC catalyst activation has been successfully applied for many years. Mori et al. [112] investigated the influence of the mechanical milling of MgO-mixed catalysts of Ru, Ni, Fe, or their combinations at temperatures ranging from 80 to 150 °C under initial pressures of 100 Torr CO$_2$ and 500 Torr H$_2$ on the hydrogenation of CO$_2$ to CH$_4$. They found that the CH$_4$ yield was related to the nature of the catalyst (Ru/MgO gave the best yield, whilst Ni/Fe showed the worst activity). Interestingly, the MC activation produced catalysts with improved CH$_4$ production rate and significantly lowered the reaction activation energy. Indeed, an apparent activation energy of 39 kJ mol^{-1} and kJ mol^{-1} was accomplished for Ni-Fe-MgO and Ru-MgO catalysts, respectively, upon the milling activation. Such values were almost halved with respect to the same systems produced without mechanical activation. Interestingly, the authors found that the catalysts submitted to further milling activation pre-treatment did not shown any improvement in the activity, suggesting that the milling during reaction played the major role. SEM revealed an increased number of larger particles at increasing milling time, due to small particle agglomeration during milling. However, no significant increase of surface area was observed. Electron probe microanalysis (EPMA) results revealed uniform distribution and aggregation of Ni, Fe, and MgO particles during milling, indicating that the activity had to be ascribed to changes in the surface structure rather than to increased surface area.

In the same years, another study reported on the influence of the MC activation of several traditional catalysts such as Ni, Zr, Zr-Ni-containing materials, Zr hydride, and different Zr-Ni hydrides employed in the hydrogenation of CO and of CO$_2$ [113]. Inactive NiO and ZrO$_2$ were also investigated for comparison purposes. In particular, the study underlined the difference between the traditional catalytic hydrogenation and the mechanically induced hydrogenation to CH$_4$. The study was carried out by employing a flow-milling vial as a catalytic reactor. The authors found that Zr and Zr-Ni hydrides were the most active in CH$_4$ production, despite the fact the hydrogen present in the mixture had a negative effect, since it overcame the CH$_4$ formation. On the contrary, Ni-based catalysts were unable to produce CH$_4$, but were active in CO disproportionation. Zirconium hydride was the only one that showed activity for CO$_2$ hydrogenation. Unexpectedly, the production of CH$_4$ was observed on NiO and ZrO_2. The differences in activity were ascribed to the structural transformations of the metal and hydride catalysts under milling, i.e., to the presence of low-coordinate centers upon MC treatment that were able to efficiently activate CO.

MC synthesis was also proven to be effective for obtaining nanophase carbides by facile ball milling of elemental iron and carbon powders at room temperature. Such catalysts were tested in the hydrogenation of CO_2 and showed superior activity and selectivity with respect to Fe/C mixtures and coarse-grained conventionally synthesized carbides [114]. It was proposed that nanocrystalline iron carbides with sizes ranging between 8 and 16 nm were the catalytically active species. The presence of defects and/or highly active sites, strained regions, and grain boundaries formed during the grinding guaranteed effective hydrocarbons chain propagation. Interestingly, also a commercial cementite powder displayed some activity upon the ball-milling activation.

Re–Co/Al_2O_3 catalysts were synthesized via ball milling by using tungsten carbide balls in a tungsten carbide container [115]. In contrast to the preparation of catalysts containing noble metal NPs, in which an oxide precursor or a metal salt are usually employed, the preparation of catalysts with transition metal involves the direct milling of the metal precursors. The obtained bimetallic Re–Co-containing catalysts were more active in the CH_4 conversion than in the CO hydrogenation, whilst the activity of the catalysts prepared by incipient wetness impregnation was exactly the opposite. Such difference was explained by assuming that crystallinity in MC prepared material was introduced upon the activation heat treatment at 650 °C. Indeed, X-ray powder diffraction indicated the presence of disordered Re–Co phases before the heat treatment.

Similar systems, such as Co–Fe and Ni catalysts supported on ZrO_2 and TiO_2, were also prepared by ball milling and showed enhanced catalytic activity in CO hydrogenation [116]. It is worth noting that upon mechanical treatment of TiO_2 in the anatase form, the authors observed the transition to brookite and rutile already at room temperature. The activity was ascribed to Co50Fe50 and Ni domains with average size of about 30 and 50 nm, distributed on the supports. XRD measurements confirmed that the mechanochemical dispersion of the Co50Fe50 and Ni phases on TiO_2 and ZrO_2 prevented the Co, Fe, and Ni oxidation by reaction with the supports. Oxidation of such species usually takes place during calcination during preparation of the catalysts.

Among the most effective and economic materials able to hydrogenate CO_2 to CH_3OH, Cu–Zn-based catalysts have been extensively studied [117–121]. Usually, commercial Cu–Zn catalysts are prepared by co-precipitation on Al_2O_3 support: such method implies several steps (precipitation, aging, filtration, drying, calcination, and reduction). Along with specific surface area and CuO and ZnO distribution, such steps have a strong influence on the activity of the final catalyst [122–125].

In particular, the distribution of CuO and ZnO has pivotal relevance during the aging, because the ion exchange between Cu^{2+} in malachite ($Cu_2CO_3(OH)_2$) and Zn^{2+} in hydrozincite ($Zn_5(OH)_6(CO_3)_2$) occurs, and the formation of zincian malachite (($Cu,Zn)_2CO_3(OH)_2$) and/or aurichalcite (($Cu,Zn)_5(OH)_6(CO_3)_2$) hydroxy carbonate precursors are extremely sensitive to pH, duration, temperature, and stirring [126–128].

To overcome such problem, very recently a new mechanical-force-driven, solid-state ion-exchange procedure has been refined [129]. It was found that the mutual substitution between Cu^{2+} and Zn^{2+} was improved with increasing mechanical ball-milling speed from 200 to 400 rpm, leading to significant enhancement of the catalytic activity. The XRD characterization shown in Figure 18 provided evidence of the presence of peaks related either to the CuO phase or the ZnO phase. Such peaks become broader and at the same time decrease in intensity upon the increase of the milling speed, indicating a decrease in crystallinity and crystal size. The absence of the peaks related to the presence of the Al_2O_3 phase was explained by the alumina change into an amorphous state after ball-milling [126,129].

The results of TEM measurements performed on the CZA400 catalyst reported in Figure 19 indicated a close cross-distribution of CuO and ZnO due to the occurrence of the solid-state ion exchange, in which ZnO NPs act as spacers, thereby avoiding the CuO agglomeration (as shown in Figure 19b).

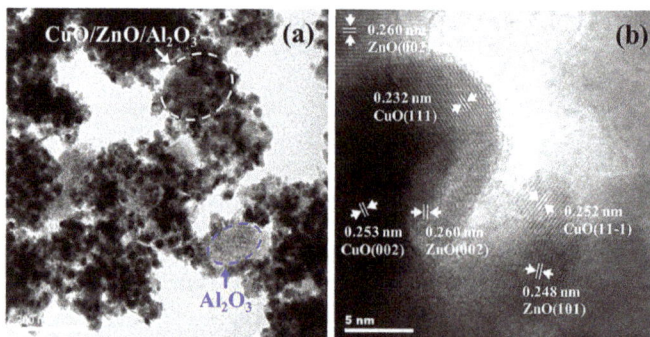

Figure 19. TEM (**a**) and HRTEM (**b**) images of CuO/ZnO/Al$_2$O$_3$ catalyst prepared at 400 rpm. Reprinted from Wu, W.; Xie, K.; Sun, D.; Li, X.; Fang, F. CuO/ZnO/Al$_2$O$_3$ Catalyst Prepared by Mechanical-Force-Driven Solid-State Ion Exchange and Its Excellent Catalytic Activity under Internal Cooling Condition. *Ind. Eng. Chem. Res.* 56, 8216–8223. Copyright (2017) American Chemical Society.

Such alternation of CuO and ZnO NPs represents the result of the complete ion exchange between Cu^{2+} and Zn^{2+} in the precursors. For the CZA400 catalyst, impressive 59.5% CO$_2$ conversion and 73.4% CH$_3$OH selectivity were obtained at 240 °C and 4 MPa, resulting in a noticeable yield to CH$_3$OH of 43.7%. Therefore, this mechanical-force-driven, solid-state ion-exchange method proved to be a valuable alternative to synthesizing CZA catalysts with improved practical application potential for chemical hydrogenation of CO$_2$ to CH$_3$OH.

In another very recent study, Au clusters were directly deposited onto a functionalized support by newly developed, milling-mediated solid reduction procedure [130]. The proposed method involved the use of HAuCl$_4$ as Au precursor, which was pre-adsorbed on the Schiff base modified silica surface by electrostatic interaction in water. After grinding the dried carrier with solid NaBH$_4$, according to Scheme 10, the Au precursor can be reduced in situ by atomic H coming from the NaBH$_4$ reducing agent in the solid state. Depending on the gold loading, highly dispersed clusters and isolated Au atoms can be obtained by solid grinding.

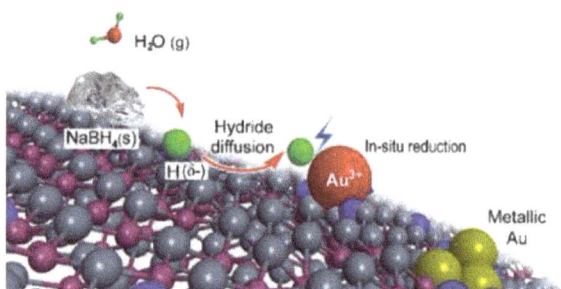

Scheme 10. Illustration of the in situ construction of ultrafine Au clusters on a nitrogenous carrier using the solid reduction method. Reprinted from Ref. [130]. Copyright 2018 with permission of Wiley.

Following such activation, it was found that the Au species were dispersed as 1.0 nm nanoclusters (Figure 20a), as determined by the aberration-corrected, HAADF-STEM (high-angle annular dark-field scanning transmission electron microscopy) analysis. At the same time, a lot of isolated Au atoms were detected, even if as a minor fraction of the total metal content. It was proposed that such Au subnanoclusters were effectively stabilized by the Schiff base groups present on the silica surface by providing strong electronic interaction. Differently, a variety of Au shapes and sizes were observed on the support (Figure 20b) when an Au-containing solution was reduced by $NaBH_4$. The obtained statistical particle size distributions were in agreement with the L_3-edge extended X-ray absorption fine structure (EXAFS, Figure 20c), revealing an increase in intensity of the peak related to the first shell, as well as in the Au coordination number when the conventional wet method was employed. In addition, a decrease in the white line intensity compared with the Au foil in the X-ray absorption near-edge structure (XANES) spectra obtained for the Au NPs (Figure 20d) is an indication that the functionalization of the silica surface with nitrogen allowed one to optimize the Au d-electron density [131].

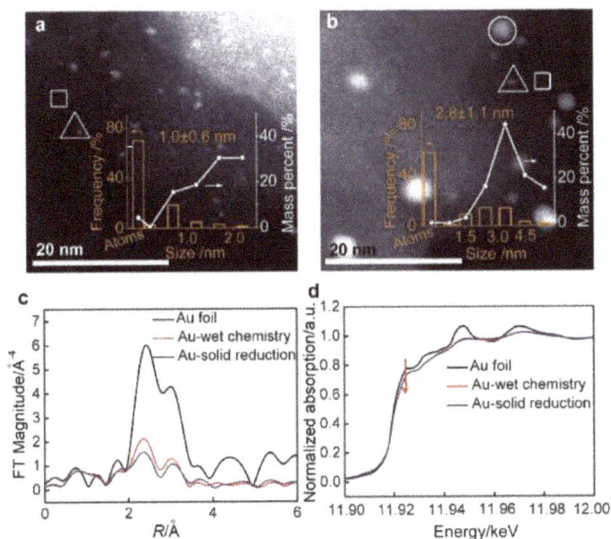

Figure 20. Aberration-corrected HAADF-STEM images of Au/SiO_2-Schiff prepared by (**a**) solid reduction method and (**b**) wet chemistry method, respectively. The NPs, subnanoclusters, and isolated Au atoms were indicated by white circles, triangles, and squares, respectively. (**c**) Fourier transform of k^3-weighted EXAFS spectra and (**d**) normalized Au L_3-edge XANES spectra of Au/SiO_2-Schiff catalysts prepared by wet chemistry and solid reduction method, respectively. Reprinted from Ref. [130]. Copyright 2018 with permission of Wiley.

These features put in evidence that the opportunity to use the solvent-free solid reduction synthesis allows the prevention of Au growth; indeed, crystal nucleation was hampered, giving rise to either subnanoclusters or single-atoms, depending on the Au content. The catalysts were tested in the hydrogenation of CO_2 to formate, and a marked dependence on the Au size, due to the different electronic structure of the metal species, was observed. The authors claimed that the milling-mediated solid reduction procedure could be considered as a general methodology to achieve basic knowledge ion the catalysis by Au, providing new insights for supported catalyst design for reactions involving the CO_2 transformation.

5. Final Remarks

The main purpose of this review was to point out the possibilities offered by microwave, ultrasound, and mechanochemical protocols in materials engineering at a nanoscale level for catalytic applications.

Summarising, concerning the synthesis of CO_2 reduction catalysts, assistance of microwave, ultrasound, and mechanochemistry allows one to obtain highly dispersed active sites and controlled particle dimension, and in most cases reinforces the active site/support interaction, giving rise to synergistic effects that help CO_2 reduction and improve catalyst stability. The procedures reported are usually simple, and require relatively short reaction times (minutes) without implying many steps. Moreover, stronger and/or more abundant basic sites occurring at the support surface favour CO_2 adsorption, a key step for its further reduction. Complex systems (hierarchical structures, composites, bi-functional catalysts, etc.) may be obtained by relatively simple methods.

Finally, a further crucial aspect concerns the effect of microwave irradiation during reaction, whilst carbon deposition usually leads to catalyst deactivation, under microwave the same carbon deposits may act as hot spots per se favouring the reaction.

Funding: This research was funded by the Università degli Studi di Torino (Progetto Ricerca Locale 2016), and Politecnico di Torino (Progetto Ricerca di base 2017).

Conflicts of Interest: The authors declare no conflict of interest.

References

1. Wang, W.; Wang, S.; Ma, X.; Gong, J. Recent advances in catalytic hydrogenation of carbon dioxide. *Chem. Soc. Rev.* **2011**, *40*. [CrossRef] [PubMed]
2. Xu, X.; Song, C.; Miller, B.G.; Scaroni, A.W. Adsorption separation of carbon dioxide from flue gas of natural gas-fired boiler by a novel nanoporous "molecular basket" adsorbent. *Fuel Process. Technol.* **2005**, *86*, 1457–1472. [CrossRef]
3. Choi, S.; Watanabe, T.; Bae, T.H.; Sholl, D.S.; Jones, C.W. Modification of the Mg/DOBDC MOF with amines to enhance CO_2 adsorption from ultradilute gases. *J. Phys. Chem. Lett.* **2012**. [CrossRef] [PubMed]
4. Kang, D.Y.; Brunelli, N.A.; Yucelen, G.I.; Venkatasubramanian, A.; Zang, J.; Leisen, J.; Hesketh, P.J.; Jones, C.W.; Nair, S. Direct synthesis of single-walled aminoaluminosilicate nanotubes with enhanced molecular adsorption selectivity. *Nat. Commun.* **2014**. [CrossRef] [PubMed]
5. Aaron, D.; Tsouris, C. Separation of CO_2 from flue gas: A review. *Sep. Sci. Technol.* **2005**, *40*, 321–348. [CrossRef]
6. Hinkov, I.; Darkrim Lamari, F.; Langlois, P.; Dicko, M.; Chilev, C.; Pentchev, I. Carbon Dioxide Capture by Adsorption (Review). *J. Chem. Technol. Metall.* **2016**, *51*, 609–626.
7. An, J.; Rosi, N.L. Tuning MOF CO_2 Adsorption Properties via Cation Exchange. *J. Am. Chem. Soc.* **2010**. [CrossRef] [PubMed]
8. Webley, P.A. Adsorption technology for CO_2 separation and capture: A perspective. *Adsorption* **2014**. [CrossRef]
9. Yu, C.H.; Huang, C.H.; Tan, C.S. A review of CO_2 capture by absorption and adsorption. *Aerosol Air Qual. Res.* **2012**. [CrossRef]
10. Gibson, J.A.A.; Mangano, E.; Shiko, E.; Greenaway, A.G.; Gromov, A.V.; Lozinska, M.M.; Friedrich, D.; Campbell, E.E.B.; Wright, P.A.; Brandani, S. Adsorption Materials and Processes for Carbon Capture from Gas-Fired Power Plants: AMPGas. *Ind. Eng. Chem. Res.* **2016**. [CrossRef]
11. Saeidi, S.; Amin, N.A.S.; Rahimpour, M.R. Hydrogenation of CO_2 to value-added products—A review and potential future developments. *J. CO_2 Util.* **2014**, *5*, 66–81. [CrossRef]
12. Milani, D.; Khalilpour, R.; Zahedi, G.; Abbas, A. A model-based analysis of CO_2 utilization in methanol synthesis plant. *J. CO_2 Util.* **2015**, *10*, 12–22. [CrossRef]
13. Wei, J.; Ge, Q.; Yao, R.; Wen, Z.; Fang, C.; Guo, L.; Xu, H.; Sun, J. Directly converting CO_2 into a gasoline fuel. *Nat. Commun.* **2017**. [CrossRef] [PubMed]
14. Gao, P.; Li, S.; Bu, X.; Dang, S.; Liu, Z.; Wang, H.; Zhong, L.; Qiu, M.; Yang, C.; Cai, J.; et al. Direct conversion of CO_2 into liquid fuels with high selectivity over a bifunctional catalyst. *Nat. Chem.* **2017**. [CrossRef] [PubMed]
15. Vesselli, E.; Monachino, E.; Rizzi, M.; Furlan, S.; Duan, X.; Dri, C.; Peronio, A.; Africh, C.; Lacovig, P.; Baldereschi, A.; et al. Steering the chemistry of carbon oxides on a NiCu catalyst. *ACS Catal.* **2013**. [CrossRef]
16. Li, K.; An, X.; Park, K.H.; Khraisheh, M.; Tang, J. A critical review of CO_2 photoconversion: Catalysts and reactors. *Catal. Today* **2014**. [CrossRef]

17. Low, J.; Cheng, B.; Yu, J. Surface modification and enhanced photocatalytic CO_2 reduction performance of TiO_2: A review. *Appl. Surf. Sci.* **2017**, *392*, 658–686. [CrossRef]

18. White, J.L.; Baruch, M.F.; Pander, J.E.; Hu, Y.; Fortmeyer, I.C.; Park, J.E.; Zhang, T.; Liao, K.; Gu, J.; Yan, Y.; et al. Light-Driven Heterogeneous Reduction of Carbon Dioxide: Photocatalysts and Photoelectrodes. *Chem. Rev.* **2015**. [CrossRef] [PubMed]

19. Nahar, S.; Zain, M.; Kadhum, A.; Hasan, H.; Hasan, M. Advances in Photocatalytic CO_2 Reduction with Water: A Review. *Materials* **2017**, *10*, 629. [CrossRef] [PubMed]

20. Zhang, H.; Kawamura, S.; Tamba, M.; Kojima, T.; Yoshiba, M.; Izumi, Y. Is water more reactive than H_2 in photocatalytic CO_2 conversion into fuels using semiconductor catalysts under high reaction pressures? *J. Catal.* **2017**. [CrossRef]

21. Kim, D.; Xie, C.; Becknell, N.; Yu, Y.; Karamad, M.; Chan, K.; Crumlin, E.J.; Nørskov, J.K.; Yang, P. Electrochemical Activation of CO_2 through Atomic Ordering Transformations of AuCu Nanoparticles. *J. Am. Chem. Soc.* **2017**. [CrossRef] [PubMed]

22. Lum, Y.; Yue, B.; Lobaccaro, P.; Bell, A.T.; Ager, J.W. Optimizing C-C Coupling on Oxide-Derived Copper Catalysts for Electrochemical CO_2 Reduction. *J. Phys. Chem. C* **2017**. [CrossRef]

23. Qiao, J.; Liu, Y.; Hong, F.; Zhang, J. A review of catalysts for the electroreduction of carbon dioxide to produce low-carbon fuels. *Chem. Soc. Rev.* **2014**. [CrossRef] [PubMed]

24. Malik, K.; Singh, S.; Basu, S.; Verma, A. Electrochemical reduction of CO_2 for synthesis of green fuel. *Wiley Interdiscip. Rev. Energy Environ.* **2017**. [CrossRef]

25. Li, C.W.; Kanan, M.W. CO_2 reduction at low overpotential on Cu electrodes resulting from the reduction of thick Cu_2O films. *J. Am. Chem. Soc.* **2012**. [CrossRef] [PubMed]

26. Huang, X.; Shen, Q.; Liu, J.; Yang, N.; Zhao, G. A CO_2 adsorption-enhanced semiconductor/metal-complex hybrid photoelectrocatalytic interface for efficient formate production. *Energy Environ. Sci.* **2016**. [CrossRef]

27. Centi, G.; Perathoner, S. Heterogeneous Catalytic Reactions with CO_2: Status and Perspectives. *Stud. Surf. Sci. Catal.* **2004**, *153*, 1–8. [CrossRef]

28. Windle, C.D.; Perutz, R.N. Advances in molecular photocatalytic and electrocatalytic CO_2 reduction. *Coord. Chem. Rev.* **2012**, *256*, 2562–2570. [CrossRef]

29. Gray, H.B. Powering the planet with solar fuel. *Nat. Chem.* **2009**. [CrossRef] [PubMed]

30. Windle, C.; Reisner, E. Heterogenised Molecular Catalysts for the Reduction of CO_2 to Fuels. *Chim. Int. J. Chem.* **2015**, *69*, 435–441. [CrossRef] [PubMed]

31. Bradford, M.C.J.; Vannice, M.A. CO_2 Reforming of CH_4. *Catal. Rev.* **1999**, *41*, 1–42. [CrossRef]

32. Ahmadi, M.; Mistry, H.; Roldan Cuenya, B. Tailoring the Catalytic Properties of Metal Nanoparticles via Support Interactions. *J. Phys. Chem. Lett.* **2016**, *7*, 3519–3533. [CrossRef] [PubMed]

33. Wang, X.-K.; Zhao, G.-X.; Huang, X.; Wang, X. Progress in the catalyst exploration for heterogeneous CO_2 reduction and utilization: A critical review. *J. Mater. Chem. A* **2017**, *5*, 21625–21649. [CrossRef]

34. Sahoo, T.R.; Armandi, M.; Arletti, R.; Piumetti, M.; Bensaid, S.; Manzoli, M.; Panda, S.R.; Bonelli, B. Pure and Fe-doped CeO_2 nanoparticles obtained by microwave assisted combustion synthesis: Physico-chemical properties ruling their catalytic activity towards CO oxidation and soot combustion. *Appl. Catal. B Environ.* **2017**, *211*. [CrossRef]

35. Devaiah, D.; Reddy, L.H.; Park, S.-E.; Reddy, B.M. Ceria-zirconia mixed oxides: Synthetic methods and applications. *Catal. Rev.* **2018**. [CrossRef]

36. Ganesh, I.; Johnson, R.; Rao, G.V.N.; Mahajan, Y.R.; Madavendra, S.S.; Reddy, B.M. Microwave-assisted combustion synthesis of nanocrystalline $MgAl_2O_4$ spinel powder. *Ceram. Int.* **2005**, *31*, 67–74. [CrossRef]

37. Ai, L.; Zeng, Y.; Jiang, J. Hierarchical porous BiOI architectures: Facile microwave nonaqueous synthesis, characterization and application in the removal of Congo red from aqueous solution. *Chem. Eng. J.* **2014**. [CrossRef]

38. Cai, W.; De La Piscina, P.R.; Toyir, J.; Homs, N. CO_2 hydrogenation to methanol over CuZnGa catalysts prepared using microwave-assisted methods. *Catal. Today* **2015**. [CrossRef]

39. Huang, C.; Mao, D.; Guo, X.; Yu, J. Microwave-Assisted Hydrothermal Synthesis of CuO-ZnO-ZrO_2 as Catalyst for Direct Synthesis of Methanol by Carbon Dioxide Hydrogenation. *Energy Technol.* **2017**, *5*, 2100–2107. [CrossRef]

40. Stangeland, K.; Kalai, D.; Yu, Z. CO_2 Methanation: The Effect of Catalysts and Reaction Conditions. *Energy Procedia* **2017**, *105*, 2022–2027. [CrossRef]

41. He, S.; Li, C.; Chen, H.; Su, D.; Zhang, B.; Cao, X.; Wang, B.; Wei, M.; Evans, D.G.; Duan, X. A Surface Defect-Promoted Ni Nanocatalyst with Simultaneously Enhanced Activity and Stability. *Chem. Mater.* **2013**, *25*, 1040–1046. [CrossRef]

42. Wu, H.C.; Chang, Y.C.; Wu, J.H.; Lin, J.H.; Lin, I.K.; Chen, C.S. Methanation of CO_2 and reverse water gas shift reactions on Ni/SiO_2 catalysts: The influence of particle size on selectivity and reaction pathway. *Catal. Sci. Technol.* **2015**, *5*, 4154–4163. [CrossRef]

43. Kesavan, J.K.; Luisetto, I.; Tuti, S.; Meneghini, C.; Iucci, G.; Battocchio, C.; Mobilio, S.; Casciardi, S.; Sisto, R. Nickel supported on YSZ: The effect of Ni particle size on the catalytic activity for CO_2 methanation. *J. CO_2 Util.* **2018**, *23*, 200–211. [CrossRef]

44. Du, G.; Lim, S.; Yang, Y.; Wang, C.; Pfefferle, L.; Haller, G.L. Methanation of carbon dioxide on Ni-incorporated MCM-41 catalysts: The influence of catalyst pretreatment and study of steady-state reaction. *J. Catal.* **2007**, *249*, 370–379. [CrossRef]

45. Hu, L.; Urakawa, A. Continuous CO_2 capture and reduction in one process: CO_2 methanation over unpromoted and promoted Ni/ZrO_2. *J. CO_2 Util.* **2018**, *25*, 323–329. [CrossRef]

46. Aziz, M.A.A.; Jalil, A.A.; Triwahyono, S.; Mukti, R.R.; Taufiq-Yap, Y.H.; Sazegar, M.R. Highly active Ni-promoted mesostructured silica nanoparticles for CO_2 methanation. *Appl. Catal. B Environ.* **2014**, *147*, 359–368. [CrossRef]

47. Song, F.; Zhong, Q.; Yu, Y.; Shi, M.; Wu, Y.; Hu, J.; Song, Y. Obtaining well-dispersed Ni/Al_2O_3 catalyst for CO_2 methanation with a microwave-assisted method. *Int. J. Hydrogen Energy* **2017**, *42*, 4174–4183. [CrossRef]

48. Bacariza, M.C.; Graça, I.; Westermann, A.; Ribeiro, M.F.; Lopes, J.M.; Henriques, C. CO_2 Hydrogenation Over Ni-Based Zeolites: Effect of Catalysts Preparation and Pre-reduction Conditions on Methanation Performance. *Top. Catal.* **2016**, *59*, 314–325. [CrossRef]

49. Dong, X.; Jin, B.; Sun, Y.; Shi, K.; Yu, L. Re-promoted Ni-Mn bifunctional catalysts prepared by microwave heating for partial methanation coupling with water gas shift under low H_2/CO conditions. *Appl. Catal. A Gen.* **2018**, *552*, 105–116. [CrossRef]

50. Fidalgo, B.; Arenillas, A.; Menéndez, J.A. Mixtures of carbon and Ni/Al_2O_3 as catalysts for the microwave-assisted CO_2 reforming of CH_4. *Fuel Process. Technol.* **2011**, *92*, 1531–1536. [CrossRef]

51. Barros, B.S.; Melo, D.M.A.; Libs, S.; Kiennemann, A. CO_2 reforming of methane over $La_2NiO_4/\alpha-Al_2O_3$ prepared by microwave assisted self-combustion method. *Appl. Catal. A Gen.* **2010**. [CrossRef]

52. Zhang, J.; Lu, S.; Su, X.; Fan, S.; Ma, Q.; Zhao, T. Selective formation of light olefins from CO_2 hydrogenation over Fe–Zn–K catalysts. *Biochem. Pharmacol.* **2015**, *12*, 95–100. [CrossRef]

53. Hou, W.; Hung, W.H.; Pavaskar, P.; Goeppert, A.; Aykol, M.; Cronin, S.B. Photocatalytic conversion of CO_2 to hydrocarbon fuels via plasmon-enhanced absorption and metallic interband transitions. *ACS Catal.* **2011**. [CrossRef]

54. Liu, E.; Qi, L.; Bian, J.; Chen, Y.; Hu, X.; Fan, J.; Liu, H.; Zhu, C.; Wang, Q. A facile strategy to fabricate plasmonic Cu modified TiO_2 nano-flower films for photocatalytic reduction of CO_2 to methanol. *Mater. Res. Bull.* **2015**. [CrossRef]

55. Shown, I.; Hsu, H.-C.; Chang, Y.-C.; Lin, C.-H.; Roy, P.K.; Ganguly, A.; Wang, C.-H.; Chang, J.-K.; Wu, C.-I.; Chen, L.-C.; et al. Highly Efficient Visible Light Photocatalytic Reduction of CO_2 to Hydrocarbon Fuels by Cu-Nanoparticle Decorated Graphene Oxide. *Nano Lett.* **2014**. [CrossRef] [PubMed]

56. An, X.; Li, K.; Tang, J. $Cu_2O/$reduced graphene oxide composites for the photocatalytic conversion of CO_2. *ChemSusChem* **2014**. [CrossRef] [PubMed]

57. Shi, G.; Yang, L.; Liu, Z.; Chen, X.; Zhou, J.; Yu, Y. Photocatalytic reduction of CO_2 to CO over copper decorated g-C3N4nanosheets with enhanced yield and selectivity. *Appl. Surf. Sci.* **2018**. [CrossRef]

58. Yu, J.; Jin, J.; Cheng, B.; Jaroniec, M. A noble metal-free reduced graphene oxide-CdS nanorod composite for the enhanced visible-light photocatalytic reduction of CO_2 to solar fuel. *J. Mater. Chem. A* **2014**. [CrossRef]

59. Fenton, T.; Gillingham, S.; Jin, T.; Li, G. Microwave-assisted deposition of a highly active cobalt catalyst on mesoporous silica for photochemical CO_2 reduction. *Dalton Trans.* **2017**, *46*, 10721–10726. [CrossRef] [PubMed]

60. Hart, E.J.; Henglein, A. Free radical and free atom reactions in the sonolysis of aqueous iodide and formate solutions. *J. Phys. Chem.* **1985**, *89*, 4342–4347. [CrossRef]

61. Luo, S.; Wu, J.; Toyir, J.; Saito, M.; Takeuchi, M.; Watanabe, T. Optimization of preparation conditions and improvement of stability of Cu/ZnO-based multicomponent catalysts for methanol synthesis from CO_2 and H_2. *Stud. Surf. Sci. Catal.* **1998**, *114*, 549–552. [CrossRef]

62. Vafaeian, Y.; Haghighi, M.; Aghamohammadi, S. Ultrasound assisted dispersion of different amount of Ni over ZSM-5 used as nanostructured catalyst for hydrogen production via CO_2 reforming of methane. *Energy Convers. Manag.* **2013**, *76*, 1093–1103. [CrossRef]

63. Gaudino, E.C.; Rinaldi, L.; Rotolo, L.; Carnaroglio, D.; Pirola, C.; Cravotto, G.; Radoiu, M.; Eynde, J.J. Vanden Heterogeneous phase microwave-assisted reactions under CO_2 or CO pressure. *Molecules* **2016**, *21*, 253. [CrossRef]

64. Park, J.Y.; Kim, M.A.; Lee, S.J.; Jung, J.; Jang, H.M.; Upare, P.P.; Hwang, Y.K.; Chang, J.-S.; Park, J.K. Preparation and characterization of carbon-encapsulated iron nanoparticles and their catalytic activity in the hydrogenation of levulinic acid. *J. Mater. Sci.* **2015**, *50*, 334–343. [CrossRef]

65. Pirola, C.; Bianchi, C.L.; Di Michele, A.; Diodati, P.; Boffito, D.; Ragaini, V. Ultrasound and microwave assisted synthesis of high loading Fe-supported Fischer-Tropsch catalysts. *Ultrason. Sonochem.* **2010**, *17*, 610–616. [CrossRef] [PubMed]

66. Chinchen, G.C.; Denny, P.J.; Jennings, J.R.; Spencer, M.S.; Waugh, K.C. Synthesis of Methanol: Part 1. Catalysts and Kinetics. *Appl. Catal.* **1988**, *36*, 1–65. [CrossRef]

67. Sun, K.; Lu, W.; Qiu, F.; Liu, S.; Xu, X. Direct synthesis of DME over bifunctional catalyst: Surface properties and catalytic performance. *Appl. Catal. A Gen.* **2003**, *252*, 243–249. [CrossRef]

68. Chen, H.Y.; Chen, L.; Lin, J.; Tan, K.L.; Li, J. Copper Sites in Copper-Exchanged ZSM-5 for CO Activation and Methanol Synthesis: XPS and FTIR Studies. *Inorg. Chem.* **1997**, *36*, 1417–1423. [CrossRef] [PubMed]

69. Ding, W.; Liu, Y.; Wang, F.; Zhou, S.; Chen, A.; Yang, Y.; Fang, W. Promoting effect of a Cu–Zn binary precursor on a ternary Cu–Zn–Al catalyst for methanol synthesis from synthesis gas. *RSC Adv.* **2014**, *4*. [CrossRef]

70. Słoczyński, J.; Grabowski, R.; Kozłowska, A.; Olszewski, P.; Stoch, J.; Skrzypek, J.; Lachowska, M. Catalytic activity of the M/(3ZnO·ZrO2) system (M = Cu, Ag, Au) in the hydrogenation of CO_2 to methanol. *Appl. Catal. A Gen.* **2004**, *278*, 11–23. [CrossRef]

71. Hansen, P.L.; Wagner, J.B.; Helveg, S.; Rostrup-Nielsen, J.R.; Clausen, B.S.; Topsøe, H. Atom-resolved imaging of dynamic shape changes in supported copper nanocrystals. *Science* **2002**, *295*, 2053–2055. [CrossRef] [PubMed]

72. Ovesen, C.V.; Clausen, B.S.; Schiøtz, J.; Stoltze, P.; Topsøe, H.; Nørskov, J.K. Kinetic Implications of Dynamical Changes in Catalyst Morphology during Methanol Synthesis over Cu/ZnO Catalysts. *J. Catal.* **1997**, *168*, 133–142. [CrossRef]

73. Choi, Y.; Futagami, K.; Fujitani, T.; Nakamura, J. The role of ZnO in Cu/ZnO methanol synthesis catalysts—Morphology effect or active site model? *Appl. Catal. A Gen.* **2001**, *208*, 163–167. [CrossRef]

74. Słoczyński, J.; Grabowski, R.; Kozłowska, A.; Olszewski, P.; Lachowska, M.; Skrzypek, J.; Stoch, J. Effect of Mg and Mn oxide additions on structural and adsorptive properties of Cu/ZnO/ZrO2 catalysts for the methanol synthesis from CO_2. *Appl. Catal. A Gen.* **2003**, *249*, 129–138. [CrossRef]

75. Omata, K.; Watanabe, Y.; Umegaki, T.; Ishiguro, G.; Yamada, M. Low-pressure DME synthesis with Cu-based hybrid catalysts using temperature-gradient reactor. *Fuel* **2002**, *81*, 1605–1609. [CrossRef]

76. Melián-Cabrera, I.; Granados, M.L.; Fierro, J.L.G. Pd-Modified Cu–Zn Catalysts for Methanol Synthesis from CO_2/H_2 Mixtures: Catalytic Structures and Performance. *J. Catal.* **2002**, *210*, 285–294. [CrossRef]

77. Pokrovski, K.A.; Rhodes, M.D.; Bell, A.T. Effects of cerium incorporation into zirconia on the activity of Cu/ZrO2 for methanol synthesis via CO hydrogenation. *J. Catal.* **2005**, *235*, 368–377. [CrossRef]

78. Znak, L.; Stołecki, K.; Zieliński, J. The effect of cerium, lanthanum and zirconium on nickel/alumina catalysts for the hydrogenation of carbon oxides. *Catal. Today* **2005**, *101*, 65–71. [CrossRef]

79. Köppel, R.A.; Stöcker, C.; Baiker, A. Copper- and Silver-Zirconia Aerogels: Preparation, Structural Properties and Catalytic Behavior in Methanol Synthesis from Carbon Dioxide. *J. Catal.* **1998**, *179*, 515–527. [CrossRef]

80. Sun, Y.; Sermon, P.A. Evidence of a metal-support interaction in sol-gel derived Cu-ZrO2 catalysts for CO hydrogenation. *Catal. Lett.* **1994**, *29*, 361–369. [CrossRef]

81. Jingfa, D.; Qi, S.; Yulong, Z.; Songying, C.; Dong, W. A novel process for preparation of a Cu/ZnO/Al2O3 ultrafine catalyst for methanol synthesis from $CO_2 + H_2$: Comparison of various preparation methods. *Appl. Catal. A Gen.* **1996**, *139*, 75–85. [CrossRef]

82. Coteron, A.; Hayhurst, A.N. Methanol synthesis by amorphous copper-based catalysts prepared by spark-erosion. *Appl. Catal. A Gen.* **1993**, *101*, 151–165. [CrossRef]

83. Arena, F.; Barbera, K.; Italiano, G.; Bonura, G.; Spadaro, L.; Frusteri, F. Synthesis, characterization and activity pattern of Cu–ZnO/ZrO$_2$ catalysts in the hydrogenation of carbon dioxide to methanol. *J. Catal.* **2007**, *249*, 185–194. [CrossRef]

84. Chen, H.-J.; Fan, C.-W.; Yu, C.-S. Analysis, synthesis, and design of a one-step dimethyl ether production via a thermodynamic approach. *Appl. Energy* **2013**, *101*, 449–456. [CrossRef]

85. Chen, H.-B.; Liao, D.-W.; Yu, L.-J.; Lin, Y.-J.; Yi, J.; Zhang, H.-B.; Tsai, K.-R. Influence of trivalent metal ions on the surface structure of a copper-based catalyst for methanol synthesis. *Appl. Surf. Sci.* **1999**, *147*, 85–93. [CrossRef]

86. Yang, R.; Yu, X.; Zhang, Y.; Li, W.; Tsubaki, N. A new method of low-temperature methanol synthesis on Cu/ZnO/Al$_2$O$_3$ catalysts from CO/CO$_2$/H$_2$. *Fuel* **2008**, *87*, 443–450. [CrossRef]

87. Tokay, K.C.; Dogu, T.; Dogu, G. Dimethyl ether synthesis over alumina based catalysts. *Chem. Eng. J.* **2012**, *184*, 278–285. [CrossRef]

88. Sai Prasad, P.S.; Bae, J.W.; Kang, S.-H.; Lee, Y.-J.; Jun, K.-W. Single-step synthesis of DME from syngas on Cu–ZnO–Al$_2$O$_3$/zeolite bifunctional catalysts: The superiority of ferrierite over the other zeolites. *Fuel Process. Technol.* **2008**, *89*, 1281–1286. [CrossRef]

89. Khoshbin, R.; Haghighi, M. Direct syngas to DME as a clean fuel: The beneficial use of ultrasound for the preparation of CuO–ZnO–Al$_2$O$_3$/HZSM-5 nanocatalyst. *Chem. Eng. Res. Des.* **2013**, *91*, 1111–1122. [CrossRef]

90. Khoshbin, R.; Haghighi, M. Direct conversion of syngas to dimethyl ether as a green fuel over ultrasound-assisted synthesized CuO–ZnO–Al$_2$O$_3$/HZSM-5 nanocatalyst: Effect of active phase ratio on physicochemical and catalytic properties at different process conditions. *Catal. Sci. Technol.* **2014**, *4*, 1779–1792. [CrossRef]

91. Allahyari, S.; Haghighi, M.; Ebadi, A.; Qavam Saeedi, H. Direct synthesis of dimethyl ether as a green fuel from syngas over nanostructured CuO–ZnO–Al$_2$O$_3$/HZSM-5 catalyst: Influence of irradiation time on nanocatalyst properties and catalytic performance. *J. Power Sources* **2014**, *272*, 929–939. [CrossRef]

92. Allahyari, S.; Haghighi, M.; Ebadi, A.; Hosseinzadeh, S. Ultrasound assisted co-precipitation of nanostructured CuO–ZnO–Al$_2$O$_3$ over HZSM-5: Effect of precursor and irradiation power on nanocatalyst properties and catalytic performance for direct syngas to DME. *Ultrason. Sonochem.* **2014**, *21*, 663–673. [CrossRef] [PubMed]

93. Prasad, K.; Pinjari, D.V.; Pandit, A.B.; Mhaske, S.T. Synthesis of zirconium dioxide by ultrasound assisted precipitation: Effect of calcination temperature. *Ultrason. Sonochem.* **2011**, *18*, 1128–1137. [CrossRef] [PubMed]

94. Zha, F.; Tian, H.; Yan, J.; Chang, Y. Multi-walled carbon nanotubes as catalyst promoter for dimethyl ether synthesis from CO$_2$ hydrogenation. *Appl. Surf. Sci.* **2013**, *285*, 945–951. [CrossRef]

95. Stark, W.J.; Stoessel, P.R.; Wohlleben, W.; Hafner, A. Industrial applications of nanoparticles. *Chem. Soc. Rev.* **2015**, *44*, 5793–5805. [CrossRef] [PubMed]

96. Behrens, S. Preparation of functional magnetic nanocomposites and hybrid materials: Recent progress and future directions. *Nanoscale* **2011**, *3*, 877–892. [CrossRef] [PubMed]

97. Singamaneni, S.; Bliznyuk, V.N.; Binek, C.; Tsymbal, E.Y. Magnetic nanoparticles: Recent advances in synthesis, self-assembly and applications. *J. Mater. Chem.* **2011**, *21*, 16819. [CrossRef]

98. Gual, A.; Godard, C.; Castillón, S.; Curulla-Ferré, D.; Claver, C. Colloidal Ru, Co and Fe-nanoparticles. Synthesis and application as nanocatalysts in the Fischer-Tropsch process. *Catal. Today* **2012**, *183*, 154–171. [CrossRef]

99. Burda, C.; Chen, X.; Narayanan, R.; El-Sayed, M.A. Chemistry and Properties of Nanocrystals of Different Shapes. *Chem. Rev.* **2005**. [CrossRef] [PubMed]

100. Vargas, E.; Romero-Sáez, M.; Denardin, J.C.; Gracia, F. The ultrasound-assisted synthesis of effective monodisperse nickel nanoparticles: Magnetic characterization and its catalytic activity in CO$_2$ methanation. *New J. Chem.* **2016**, *40*, 7307–7310. [CrossRef]

101. Yamasaki, M.; Habazaki, H.; Asami, K.; Izumiya, K.; Hashimoto, K. Effect of tetragonal ZrO$_2$ on the catalytic activity of Ni/ZrO$_2$ catalyst prepared from amorphous Ni–Zr alloys. *Catal. Commun.* **2006**, *7*, 24–28. [CrossRef]

102. Yamasaki, M.; Komori, M.; Akiyama, E.; Habazaki, H.; Kawashima, A.; Asami, K.; Hashimoto, K. CO_2 methanation catalysts prepared from amorphous Ni–Zr–Sm and Ni–Zr–misch metal alloy precursors. *Mater. Sci. Eng. A* **1999**, *267*, 220–226. [CrossRef]

103. Men, Y.; Kolb, G.; Zapf, R.; Löwe, H. Selective methanation of carbon oxides in a microchannel reactor—Primary screening and impact of gas additives. *Catal. Today* **2007**, *125*, 81–87. [CrossRef]

104. Sominski, E.; Gedanken, A.; Perkas, N.; Buchkremer, H.P.; Menzler, N.H.; Zhang, L.Z.; Yu, J.C. The sonochemical preparation of a mesoporous NiO/yttria stabilized zirconia composite. *Microporous Mesoporous Mater.* **2003**, *60*, 91–97. [CrossRef]

105. Perkas, N.; Amirian, G.; Zhong, Z.; Teo, J.; Gofer, Y.; Gedanken, A. Methanation of Carbon Dioxide on Ni Catalysts on Mesoporous ZrO_2 Doped with Rare Earth Oxides. *Catal. Lett.* **2009**, *130*, 455–462. [CrossRef]

106. Daroughegi, R.; Meshkani, F.; Rezaei, M. Enhanced activity of CO_2 methanation over mesoporous nanocrystalline Ni–Al_2O_3 catalysts prepared by ultrasound-assisted co-precipitation method. *Int. J. Hydrogen Energy* **2017**, *42*, 15115–15125. [CrossRef]

107. Kruatim, J.; Jantasee, S.; Jongsomjit, B. Improvement of cobalt dispersion on Co/SBA-15 and Co/SBA-16 catalysts by ultrasound and vacuum treatments during post-impregnation step. *Eng. J.* **2017**, *21*, 17–28. [CrossRef]

108. Zhang, Y.; Zhang, G.; Wang, L.; Xu, Y.; Sun, Y. Selective methanation of carbon monoxide over Ru-based catalysts in H_2-rich gases. *J. Ind. Eng. Chem.* **2012**, *18*, 1590–1597. [CrossRef]

109. Xu, C.; De, S.; Balu, A.M.; Ojeda, M.; Luque, R. Mechanochemical synthesis of advanced nanomaterials for catalytic applications. *Chem. Commun.* **2015**, *51*, 6698–6713. [CrossRef] [PubMed]

110. Martins, M.A.P.; Frizzo, C.P.; Moreira, D.N.; Buriol, L.; Machado, P. Solvent-Free Heterocyclic Synthesis. *Chem. Rev.* **2009**, *109*, 4140–4182. [CrossRef] [PubMed]

111. Mitchenko, S.A. Mechanochemistry in heterogeneous catalysis. *Theor. Exp. Chem.* **2007**, *43*, 211–228. [CrossRef]

112. Mori, S.; Xu, W.-C.; Ishizuki, T.; Ogasawara, N.; Imai, J.; Kobayashi, K. Mechanochemical activation of catalysts for CO_2 methanation. *Appl. Catal. A Gen.* **1996**, *137*, 255–268. [CrossRef]

113. Morozova, O.S.; Streletskii, A.N.; Berestetskaya, I.V.; Borunova, A.B. Co and Co_2 hydrogenation under mechanochemical treatment. *Catal. Today* **1997**, *38*, 107–113. [CrossRef]

114. Trovarelli, A.; Matteazzi, P.; Dolcetti, G.; Lutman, A.; Miani, F. Nanophase iron carbides as catalysts for carbon dioxide hydrogenation. *Appl. Catal. A Gen.* **1993**, *95*, L9–L13. [CrossRef]

115. Guczi, L.; Takács, L.; Stefler, G.; Koppány, Z.; Borkó, L. Re–Co/Al_2O_3 bimetallic catalysts prepared by mechanical treatment: CO hydrogenation and CH_4 conversion. *Catal. Today* **2002**, *77*, 237–243. [CrossRef]

116. Delogu, F.; Mulas, G.; Garroni, S. Hydrogenation of carbon monoxide under mechanical activation conditions. *Appl. Catal. A Gen.* **2009**, *366*, 201–205. [CrossRef]

117. Meunier, F.C. Mixing Copper Nanoparticles and ZnO Nanocrystals: A Route towards Understanding the Hydrogenation of CO_2 to Methanol? *Angew. Chem. Int. Ed.* **2011**, *50*, 4053–4054. [CrossRef] [PubMed]

118. Gao, P.; Li, F.; Zhan, H.; Zhao, N.; Xiao, F.; Wei, W.; Zhong, L.; Wang, H.; Sun, Y. Influence of Zr on the performance of Cu/Zn/Al/Zr catalysts via hydrotalcite-like precursors for CO_2 hydrogenation to methanol. *J. Catal.* **2013**, *298*, 51–60. [CrossRef]

119. Liao, F.; Huang, Y.; Ge, J.; Zheng, W.; Tedsree, K.; Collier, P.; Hong, X.; Tsang, S.C. Morphology-Dependent Interactions of ZnO with Cu Nanoparticles at the Materials' Interface in Selective Hydrogenation of CO_2 to CH_3OH. *Angew. Chem. Int. Ed.* **2011**, *50*, 2162–2165. [CrossRef] [PubMed]

120. Liu, X.-M.; Lu, G.Q.; Yan, Z.-F.; Beltramini, J. Recent Advances in Catalysts for Methanol Synthesis via Hydrogenation of CO and CO_2. *Ind. Eng. Chem. Res.* **2003**. [CrossRef]

121. Raudaskoski, R.; Turpeinen, E.; Lenkkeri, R.; Pongrácz, E.; Keiski, R.L. Catalytic activation of CO_2: Use of secondary CO_2 for the production of synthesis gas and for methanol synthesis over copper-based zirconia-containing catalysts. *Catal. Today* **2009**, *144*, 318–323. [CrossRef]

122. Fujitani, T.; Nakamura, J. The effect of ZnO in methanol synthesis catalysts on Cu dispersion and the specific activity. *Catal. Lett.* **1998**, *56*, 119–124. [CrossRef]

123. Behrens, M.; Zander, S.; Kurr, P.; Jacobsen, N.; Senker, J.; Koch, G.; Ressler, T.; Fischer, R.W.; Schlögl, R. Performance Improvement of Nanocatalysts by Promoter-Induced Defects in the Support Material: Methanol Synthesis over Cu/ZnO:Al. *J. Am. Chem. Soc.* **2013**, *135*, 6061–6068. [CrossRef] [PubMed]

124. Arena, F.; Italiano, G.; Barbera, K.; Bonura, G.; Spadaro, L.; Frusteri, F. Basic evidences for methanol-synthesis catalyst design. *Catal. Today* **2009**, *143*, 80–85. [CrossRef]
125. Behrens, M. Meso- and nano-structuring of industrial Cu/ZnO/(Al$_2$O$_3$) catalysts. *J. Catal.* **2009**, *267*, 24–29. [CrossRef]
126. Baltes, C.; Vukojević, S.; Schüth, F. Correlations between synthesis, precursor, and catalyst structure and activity of a large set of CuO/ZnO/Al$_2$O$_3$ catalysts for methanol synthesis. *J. Catal.* **2008**, *258*, 334–344. [CrossRef]
127. Behrens, M.; Schlögl, R. How to Prepare a Good Cu/ZnO Catalyst or the Role of Solid State Chemistry for the Synthesis of Nanostructured Catalysts. *Z. Anorg. Allg. Chem.* **2013**, *639*, 2683–2695. [CrossRef]
128. Li, Z.; Yan, S.; Fan, H. Enhancement of stability and activity of Cu/ZnO/Al$_2$O$_3$ catalysts by microwave irradiation for liquid phase methanol synthesis. *Fuel* **2013**, *106*, 178–186. [CrossRef]
129. Wu, W.; Xie, K.; Sun, D.; Li, X.; Fang, F. CuO/ZnO/Al$_2$O$_3$ Catalyst Prepared by Mechanical-Force-Driven Solid-State Ion Exchange and Its Excellent Catalytic Activity under Internal Cooling Condition. *Ind. Eng. Chem. Res.* **2017**, *56*, 8216–8223. [CrossRef]
130. Liu, Q.; Wang, X.; Ren, Y.; Yang, X.; Wu, Z.; Liu, X.; Li, L.; Miao, S.; Su, Y.; Li, Y.; et al. Synthesis of Subnanometer-Sized Gold Clusters by a Simple Milling-Mediated Solid Reduction Method. *Chin. J. Chem.* **2018**, *36*, 329–332. [CrossRef]
131. Liu, F.; Zhang, P. Tailoring the local structure and electronic property of AuPd nanoparticles by selecting capping molecules. *Appl. Phys. Lett.* **2010**, *96*, 043105. [CrossRef]

catalysts

MDPI

Article

High Pressure Photoreduction of CO_2: Effect of Catalyst Formulation, Hole Scavenger Addition and Operating Conditions

Elnaz Bahadori [1,2], **Antonio Tripodi** [1], **Alberto Villa** [1], **Carlo Pirola** [1], **Laura Prati** [1], **Gianguido Ramis** [2,*] **and Ilenia Rossetti** [1,*]

1   Dip. Chimica, Università degli Studi di Milano, INSTM Unit Milano-Università and CNR-ISTM, via C. Golgi, 19, I-20133 Milano, Italy; elnaz.bahadori@unimi.it (E.B.); antonio.tripodi@unimi.it (A.T.); alberto.villa@unimi.it (A.V.); carlo.pirola@unimi.it (C.P.); laura.prati@unimi.it (L.P.)
2   Dip. di Ingegneria Civile, Chimica e Ambientale, Università degli Studi di Genova and INSTM Unit Genova, via all'Opera Pia 15A, I-16100 Genoa, Italy
*   Correspondence: gianguidoramis@unige.it (G.R.); ilenia.rossetti@unimi.it (I.R.);
    Tel.: +39-010-3536027 (G.R.); +39-02-50314059 (I.R.); Fax: +39-010-3530628 (G.R.); +39-02-50314300 (I.R.)

Received: 15 August 2018; Accepted: 28 September 2018; Published: 30 September 2018

Abstract: The photoreduction of CO_2 is an intriguing process which allows the synthesis of fuels and chemicals. One of the limitations for CO_2 photoreduction in the liquid phase is its low solubility in water. This point has been here addressed by designing a fully innovative pressurized photoreactor, allowing operation up to 20 bar and applied to improve the productivity of this very challenging process. The photoreduction of CO_2 in the liquid phase was performed using commercial TiO_2 (Evonink P25), TiO_2 obtained by flame spray pyrolysis (FSP) and gold doped P25 (0.2 wt% Au-P25) in the presence of Na_2SO_3 as hole scavenger (HS). The different reaction parameters (catalyst concentration, pH and amount of HS) have been addressed. The products in liquid phase were mainly formic acid and formaldehyde. Moreover, for longer reaction time and with total consumption of HS, gas phase products formed (H_2 and CO) after accumulation of significant number of organic compounds in the liquid phase, due to their consecutive photoreforming. Enhanced CO_2 solubility in water was achieved by adding a base (pH = 12–14). In basic environment, CO_2 formed carbonates which further reduced to formaldehyde and formic acid and consequently formed $CO/CO_2 + H_2$ in the gas phase through photoreforming. The deposition of small Au nanoparticles (3–5 nm) (NPs) onto TiO_2 was found to quantitatively influence the products distribution and increase the selectivity towards gas phase products. Significant energy storage in form of different products has been achieved with respect to literature results.

Keywords: CO_2 reduction; photoreduction; Titania; photocatalysis; high pressure photocatalysis

1. Introduction

Carbon dioxide (CO_2) is one of the most important greenhouse gases emitted in the atmosphere and one of the main sources of global warming. According to the Intergovernmental Panel on Climate Change (IPCC 2001) Earth surface temperature has risen by approximately 0.6 °C in the past century. Accordingly, the Paris Agreement within 195 nations reached at COP21 in December 2015 was a major milestone capping more than two decades of global negotiations aimed at averting dangerous climate change and investments towards a low carbon, resilient and sustainable future.

Several studies have been focused on the activation of the very stable CO_2 molecule coming from carbon-capture and storage technologies (CCS) and converting it into useful chemicals for its valorisation [1]. The most interesting methods attempt the conversion of CO_2 into other useful compounds, for example, regenerated fuels or chemicals, through chemical reactions [2], catalytic [3] and photocatalytic processes [4].

CO_2 is a relatively inert and stable compound, therefore its reduction by H_2O to form hydrocarbons is an "uphill" ($\Delta G > 0$) and strongly endothermic process, requiring a considerable amount of energy [5]. Photocatalysis seems to represent a valid and green method, which may exploit solar energy for the sustainable reduction of CO_2 using H_2O as both an electron donor and a proton source at a low temperature and its conversion to useful products such as carbon monoxide (CO), formate, methanol, methane and oxygen (O_2) (Scheme 1) [6].

There are three main factors which play an important role in the photocatalytic process: solar light harvesting, separation of the photoproduced charges and surface reaction. Significant improvements have been achieved for optimization of the first 2 steps since they are based on the same issues as the widely studied for other photocatalytic applications, for example, solar driven water splitting. The major difference is the surface reaction of charge carriers [7,8]. In case of CO_2 photoreduction, the surface reaction is very challenging due to severe competition with hydrogen evolution reaction (HER) in the presence of water, which is more abundant and preferentially adsorbed onto the catalyst surfaces than CO_2 [9,10]. Hence, design and fabrication of efficient photocatalysts for CO_2 reduction is the aim of several studies [7,10–13].

TiO_2, as a low-cost semiconductor, resistant to photo-corrosion, has been widely studied for the adsorption, photoinduced activation and reduction of CO_2 [14–17]. He et al. proved that the anatase (101) facet played a critical role in CO_2 adsorption and assisting the electron transfer from the surface of TiO_2 to CO_2 in the photoreduction process [18,19]. Besides, TiO_2, shows favourable behaviour toward generating and separating electron–hole pairs during photoexcitation [16]. However, in order to improve the catalyst efficiency, decreasing the band gap and the fast recombination rate of holes and electrons generated during the irradiation process are the main concerns. To overcome this problem, various approaches have been developed: (i) noble metals addition to TiO_2 acting as electron sinks (e.g., Au, Cu, Ag, Pd) [20,21], (ii) the use of organic or inorganic hole scavengers (HS) to donate electrons to the valence band of the semiconductor preventing the accumulation of holes [11]. Even though the use of HS has been shown to enhance the rate of the photocatalytic process, the by-products forming in their presence have to be also considered [22]. Sodium sulphite was chosen because of its ability to be oxidized into sulphate by the photogenerated holes and because it is considered as a non-harmful, widely abundant compound [23].

The photoreduction reaction involves multiple proton-coupled electron transfer reactions and can lead to the formation of many different products, either in liquid phase: HCOOH, HCHO, CH_3OH or in the gas phase: H_2, CO, CH_4, depending on the reaction pathways, which makes this process rather complex (Scheme 1). A comparative study has been also carried out between the reaction in gas or liquid phase, the latter being the most promising [24] and leading to a promising route for the storage of solar energy in form of organic molecules [25]. We have already reported an innovative high pressure photoreactor, operating up to 20 bar [26,27] to successfully improve CO_2 solubility. In that investigation, we have demonstrated that the solubility of CO_2 in water is greatly enhanced at increasing pressure. Furthermore, a significant increase of activity can be achieved by increasing temperature, likely speeding up the dark steps of the reaction. Of course, the increase of temperature decreases the concentration of dissolved CO_2 but the effect of pressure is by far more significant, so that operation at 7 bar, 80 °C leads to ca. 0.1 mol% CO_2 in liquid phase, whereas in ambient conditions the value is more than ca. 5 times lower [24]. According to the higher solubility, the increase of pressure boosted the productivity of liquid phase products (HCOOH, HCHO and CH_3OH, depending on catalyst formulation and conditions). Too high pressure depressed the productivity in the gas phase. Thus, operation at intermediate pressure (ca. 7 bar) allows to evidence the effect of the operating parameter on the whole spectrum of products.

This work reports a comprehensive study on CO_2 photoreduction according to several variables. The reaction pathways and, thus, the control on products formation can be tuned by acting on different reaction parameters (e.g., pH, catalyst concentration and the amount of HS). Furthermore, improving the light harvesting capacity of TiO_2 by doping with Au nanoparticles, which also act as electron sinks, affected both productivity and products distribution. The photocatalytic activity of TiO_2 samples obtained by different preparation routes has been also investigated.

The specific configuration of the photoreactor suites the appropriate light distribution in the whole area. The very high productivity of H_2 and HCOOH even with bare P25 photocatalyst, with respect to previous studies on TiO_2 base photocatalyst, confirms the efficiency of our photoreactor.

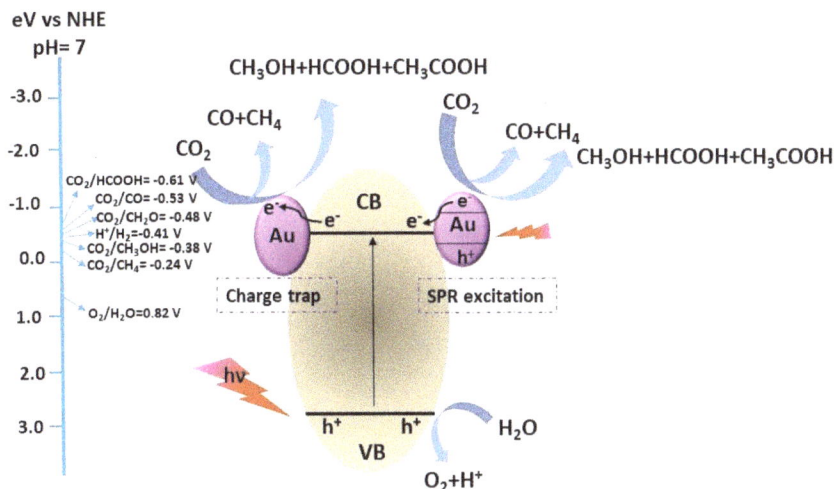

Scheme 1. Schematic illustration of different possible photocatalytic products formation during CO_2 photoreduction with H_2O over a heterogeneous photocatalyst and standard reduction potentials (V).

2. Results and Discussion

2.1. Materials Characterization

The XRD pattern (Figure 1) of TiO_2 sample obtained by flame pyrolysis shows a mixture of the crystalline phases of anatase and rutile with similar composition and particle size with respect to P25 samples (Table 1). All the diffraction features were identified by comparison with the standard JCPDS spectrum of rutile (file 88-1175) and anatase (file 84-1286) [28]. The phase composition and the average particle of each sample have been estimated from the intensity ratio between the reflection of anatase and rutile planes at (101) and (110) respectively (Table 1) [29]. The particle size of TiO_2 samples has been calculated by using the Scherrer's equation [30].

The BET SSA (Brunauer-Emmett-Teller Specific Surface Area) and pore volume have been determined based on N_2 adsorption/desorption isotherms, collected at $-196\,°C$ for P25 and FSP samples, previously outgassed at $150\,°C$ for 4 h (Figure 2). Micropore volume was calculated according to the *t*-plot method (Table 1). Both P25 and FSP samples show a type II isotherm with H1 hysteresis loop, representing the agglomerates or spherical particles arranged uniformly with high pore size uniformity and facile pore connectivity [31]. FSP samples, however, show higher surface area and pore volume with respect to P25, which may positively affect its catalytic performance. The surface area was not depressed by Au addition, given its very low amount. On the contrary, a slight increase of surface area even occurred during the chemical treatment of deposition, with a parallel increase of the total porosity of the sample with loss of microporosity.

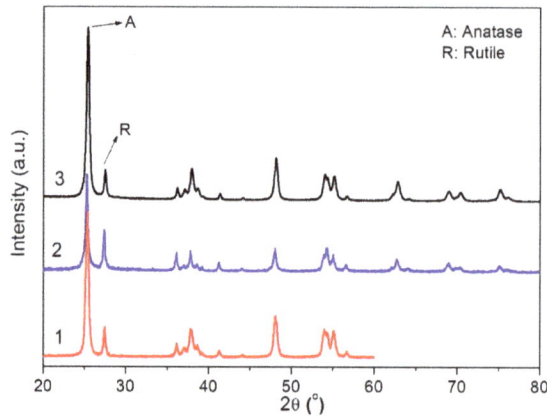

Figure 1. XRD patterns of P25 (1), 0.2 wt% Au-P25 (2) and FSP (3). A and R stand for anatase and rutile phases, respectively.

Table 1. Some relevant properties of the samples, as derived by N_2 sorption isotherms at -196 °C, XRD patterns and Band gap calculation from DR UV-Vis data elaborated according to Tauc plots.

Sample	P25	FSP	0.2 wt% Au-P25
Anatase/Rutile (%)	78/22	69/31	78/22
Crystallite size (nm) [a]	15	20	15
BET Surface area ($m^2 \cdot g^{-1}$) [b]	45	67	55
Total pore volume ($cm^3 \cdot g^{-1}$) [c]	0.12	0.14	0.27
t-Plot micropore volume ($cm^3 \cdot g^{-1}$) [c]	0.01	0.02	0.005
BJH Adsorption average pore width (nm)	22	20	31
Band Gap energy (eV) [d]	3.36	3.36	3.17

[a] Crystallite size quantification from XRD data through the Scherrer equation. [b] as calculated from N_2 adsorption/desorption isotherms, collected at -196 °C. [c] as calculated by applying the t-plot. [d] as calculated by the Tauc equation to DR-UV-Vis spectra.

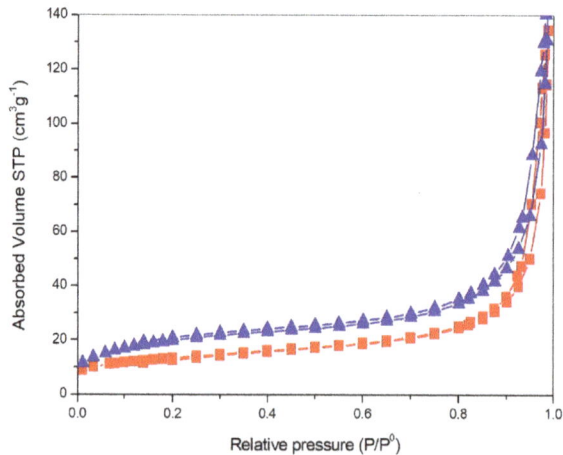

Figure 2. N_2 adsorption/desorption isotherms collected at -196 °C over samples outgassed overnight at 150 °C, P25 (squares), FSP (triangles).

According to UV absorption spectra (Figure 3a), both TiO_2 and Au-TiO_2 samples show an intense absorption in the spectral range between 240–380 nm, due to electron transfer from the 2p valence band orbital of O to the 3d conduction band orbital of Ti [32,33]. The spectra of un-doped TiO_2 samples show the cut-off at shorter wavelengths, with respect to the doped samples. The main reason for the observed bathochromic shift in transition and the visible light absorption is due to changing of the energy levels of the semiconductor band gap through a charge transfer between the metal conduction band and the valence band or the d–d transition in the crystal field [30].

In addition, the Au-TiO$_2$ sample exhibits significantly enhanced light absorption in the visible region showing a broad band located between 450 and 600 nm typical of the Surface Plasmonic Resonance (SPR) of Au nanoparticles (NPs) (inset of Figure 3a). The broad visible light absorption range is possibly due to wide size distribution of Au-NPs and the maximum of the SPR band (λ_{max}) intensity is mainly related to the size and content of Au particles.

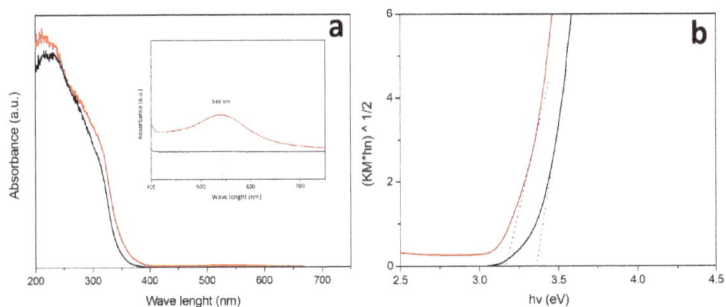

Figure 3. DR UV-Vis spectra (**a**) and corresponding Tauc plots (**b**) of P25 (black curve) and promoted with Au (0.2 wt%; red curve).

The optical band gap energy E_g was determined according to the Tauc equation [34].

According to the E_g calculations (Figure 3b and Table 1) by promoting the TiO$_2$ samples with Au, the absorption has been extended to longer wavelengths and the band gap energy reduced [32,35].

Au particle size distribution was determined from HRTEM and STEM images. Representative images are reported for 0.2 wt% Au-P25 with the respective histogram in Figure 4, revealing very small particles with a fairly narrow size distribution. Mean Au particle size was 3.6 nm.

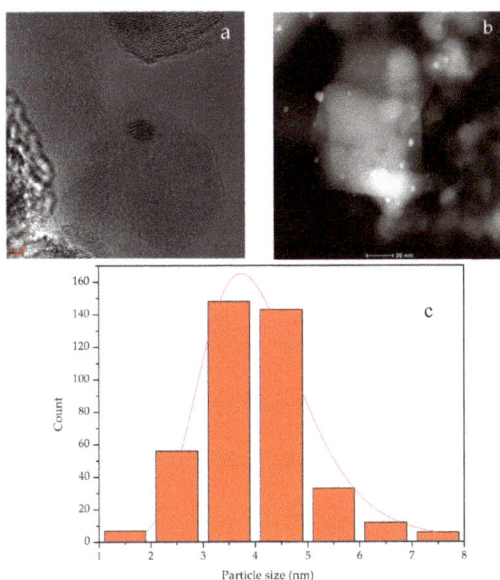

Figure 4. Representative HRTEM (**a**) and STEM (**b**) images; particle size distribution (**c**). Sample 0.2 wt% Au-P25.

2.2. CO_2 Photo-Reduction

2.2.1. Effect of pH

The photoreduction of CO_2 may lead to a broad spectrum of products depending on photocatalyst formulation and reaction conditions, due to occurrence of many parallel and consecutive reaction steps [26,27,36].

The productivity and selectivity of products on P25 has been studied at pH 7.5 and 14 in the presence of 1.66 g L^{-1} HS (Figure 5). The productivity of the main products, HCOOH and H_2, increased in basic pH in fair agreement with previous observations [26,37]. Increasing the pH improves the CO_2 solubility by forming CO_3^- or HCO_3^-, which further reduced to HCOOH or HCHO in a series of subsequent reactions (Scheme 2). Furthermore, the formed liquid products may evolve to gas phase products (H_2 and CO) due to the consecutive step of photo-reforming (Scheme 2) [26,37]. According to Ao et al. [38] in studies at basic pH, the back-oxidation of HCHO to HCOOH is more likely.

No methane formation has been observed, since P25 as a photocatalyst is likely to produce CO or HCOOH and is not likely to generate highly reduced hydrocarbons [39–42]. However, CO can be the precursor of methane formation following an alternative hydrogenation pathway [43].

Blank tests with the catalyst without irradiation and by irradiating without any catalyst revealed undetectable productivity to any species. A further photoreduction tests without pre-saturation with CO_2, pressurizing with N_2, led to nil concentration of organic products in liquid or gas phase, with a hydrogen productivity of 134.6 mmol H_2 kg_{cat}^{-1} h^{-1} due to the contribution of the direct water splitting, promoted by the presence of the hole scavenger. Therefore, it should be firmly remarked that the products formed in liquid phase are genuinely due to the reduction of CO_2.

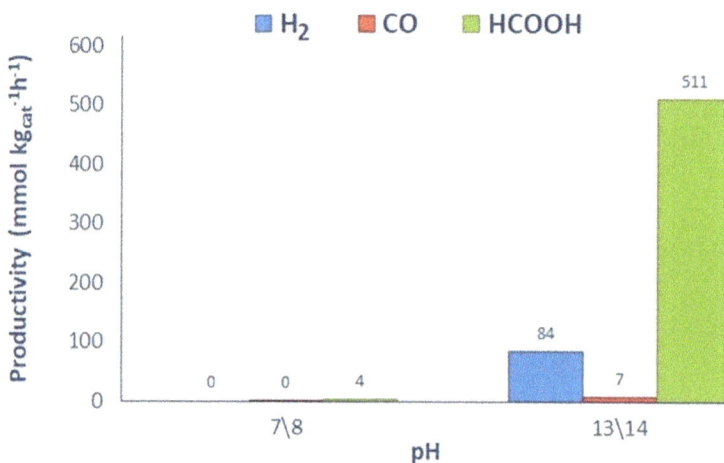

Figure 5. Influence of pH over productivity. Reaction conditions: 0.5 g L^{-1} of P25, 1.66 g L^{-1} HS.

$$*CO_3^{\cdot\cdot} \underset{\longleftarrow}{\overset{-e^-}{\longrightarrow}} *CO_3^- \underset{\longleftarrow}{\overset{-e^-}{\longrightarrow}} *CO_3$$

$$*CO + *O_2$$

$$HCOOH \underset{\overset{-2OH^- + 2e^- + H_2O}{\longleftarrow}}{} HCHO \overset{-2OH^- + 2e^- +2H_2O}{\longrightarrow} CH_3OH$$

$-2OH^- + 2e^- +2H_2O$

$$CO/CO_2 +H_2$$

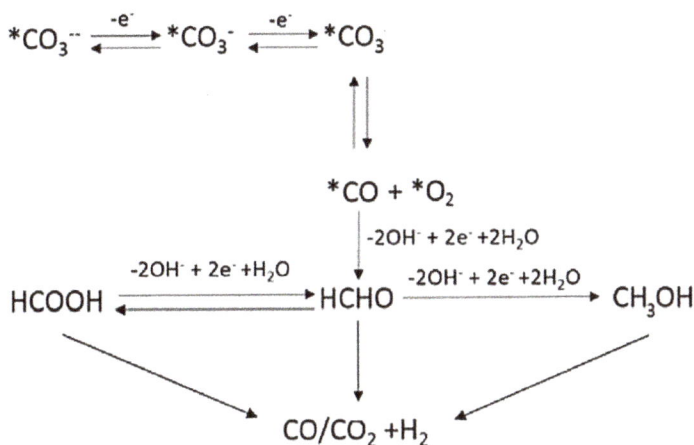

Scheme 2. Consecutive pathways for CO_2 photoreduction and photoreforming occurring at basic pH [37].

2.2.2. Effect of Catalyst Amount

Optimization of catalyst amount has been performed halving progressively the catalyst concentration (0.5, 0.25, 0.125, 0.064 and 0.031 g L^{-1}) in the photoreactor by using the bare P25 catalyst. According to Figure 6, lower catalyst concentration increased productivity mainly due to better light distribution through the whole reactor. The productivity of the gas phase products (H_2 and CO) and of HCOOH, either normalized per mass of catalyst (Figure 6a) or not (Figure 6b) allowed to assess the best catalyst concentration in the slurry. 0.031 g L^{-1} of P25 returned the highest amounts of H_2 and HCOOH (Figure 6) per mass of catalyst. All the productivities decreased when increasing catalyst mass, as quite obvious due to the normalization on catalyst mass itself. When looking at the data without normalising against catalyst mass, the highest yield in HCOOH was obtained with the highest catalyst amount and progressively decreased with decreasing catalyst concentration. In a symmetric way, hydrogen and CO yields decreased progressively with increasing catalyst concentration (Figure 6b). Therefore, this parameter can be chosen to tune the process towards the maximisation of liquid or gas phase products, depending on process goals.

0.031 g L^{-1} of catalyst was here taken as reference for further testing, so focusing on the highest gas phase productivity. Indeed, looking at the products distribution and intending this process as a mean to store solar energy by turning a waste greenhouse gas into useful compounds, we calculated the amount of energy stored in chemical form considering the different products we have obtained. We have taken as basis for calculation the enthalpy of combustion of HCOOH, H_2 and CO and made the calculation for the two extreme cases of catalyst concentration 0.031 and 0.5 g L^{-1} (Table 2). The amount of energy that is stored is slightly higher for the highest catalyst concentration and increases progressively with this parameter. However, the form of storage is different, as well as the easiness of separation and exploitation, which is likely better in the case of gas products than for the diluted liquid product. It is therefore possible to operate obtaining the highest yield of gaseous products, at low catalyst concentration, or to increase the liquid product yield with higher catalyst amount. In the following, we selected to use the lowest catalyst concentration, since it is more amenable for scale up and it leads to a balanced production of gas and liquid phase compounds, that allow to highlight the effect of the other operating parameters on reactivity.

Table 2. Amount of stored energy (kJ h^{-1}) in the form of different reaction products.

		Chemically Stored Energy (kJ h^{-1})				
		HS = 1.66 g L^{-1}				HS = 6.68 g L^{-1}
		P25		FSP	0.2 wt% Au/P25	P25
	Heat comb. (kJ mol^{-1})	Cat = 0.031 g L^{-1}	Cat = 0.5 g L^{-1}	Cat = 0.031 g L^{-1}	Cat = 0.031 g L^{-1}	Cat = 0.031 g L^{-1}
HCOOH	254	0.036	0.066	0.069	0.066	0.372
H$_2$	286	0.034	0.011	0.018	0.021	0.005
CO	283	0.003	0.001	0.002	0.003	0.0003
Total	-	0.073	0.079	0.089	0.090	0.377

Based on the maximum amount of energy stored as calculated in Table 2 and on the measured irradiance in the UVA region (104 W/m^2), we calculated approximately 2–3% energy storage efficiency.

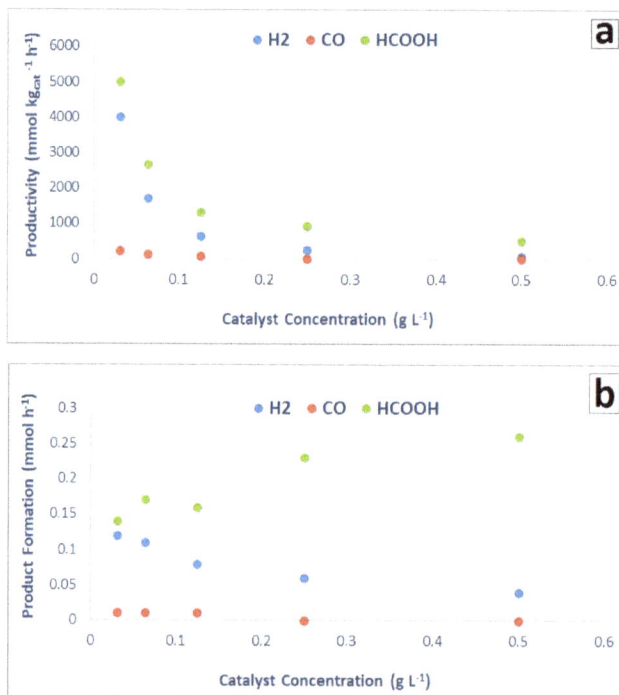

Figure 6. Effect of catalyst concentration on productivity (**a**) and absolute product formation (**b**) with the P25 as a catalyst (1.66 g L^{-1} HS and at pH = 13/14).

2.2.3. Effect of the Hole Scavenger (HS)

The efficiency of the photocatalyst is limited by the slow charge transfer, subsequent reactions of the photo-excited holes and the high charge recombination rates. In fact, consumption of conduction band electrons must be efficiently balanced by holes reduction. This process occurs in the presence of electron donor species; otherwise reaction rate is highly depressed.

Sodium sulphite (Na$_2$SO$_3$) was chosen as inorganic HS, added in different concentration (ca. 1.66, 3.34, 6.68 g L^{-1}) and its consumption was determined after reaction by iodometric titration. Negligible productivity has been observed without HS addition. Sodium sulphite is an inorganic and non-competing species, with a high performance in photocatalytic reactions [23]. Moreover, sodium sulphite can be industrially employed due to its low cost.

Figure 7 reports that increasing the HS concentration from 1.66 to 6.68 g L^{-1} (4-fold increase) increases the HCOOH productivity up to 9 times, whereas clearly decreasing the formation of gas products (H$_2$ and CO). The results of our previous studies already demonstrated that the formation of the gas phase products (H$_2$, CO, CH$_4$) starts after the consumption of the HS. Indeed, in absence of the sulphite the organic compounds accumulated in the reaction medium start acting as hole scavengers themselves through a consecutive photoreforming path (Scheme 2). Sulphites titration confirms the total consumption in 24 h of the base-case concentration 1.66 g L^{-1} (Figure 8). However, 24 h of reaction time were not enough for the total consumption of the HS when loaded in higher quantity (3.34 and 6.68 g L^{-1}), which in turn inhibits the formation of gaseous products and favours the formation of HCOOH (Figure 8). This study supports the above proposed mechanism of the reaction and the role of HS (reaction time) on the selectivity to the different products.

Moreover, according to Table 2, the boosted productivity to HCOOH, which is unprecedented in previous reports on this reaction, allows to tune also HS concentration, in addition to catalyst one, to address the reaction towards the desired products. It may be noticed, indeed, that the test with the highest HS concentration led to the highest amount of stored energy, which increased by one order of magnitude the stored energy amount even by increasing its concentration by a factor of 4, only. The choice of its use should be determined on the basis of the desired reaction path. The increase of HS makes photoreduction essentially more effective when leading to HCOOH. Its further transformation to H$_2$ is inhibited until the complete consumption of the HS, leading to liquid phase products, only.

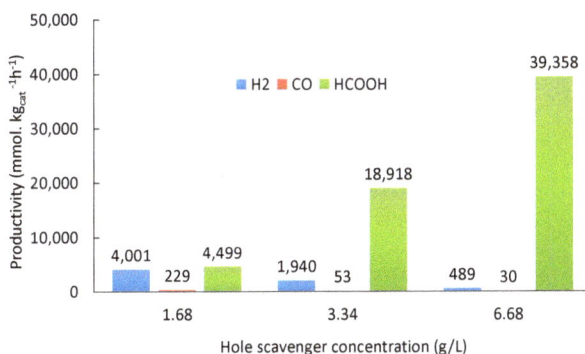

Figure 7. Effect of HS concentration on productivity. Reaction time: 24 h. P25 as a photocatalyst (0.031 g L^{-1}) and at pH = 13/14.

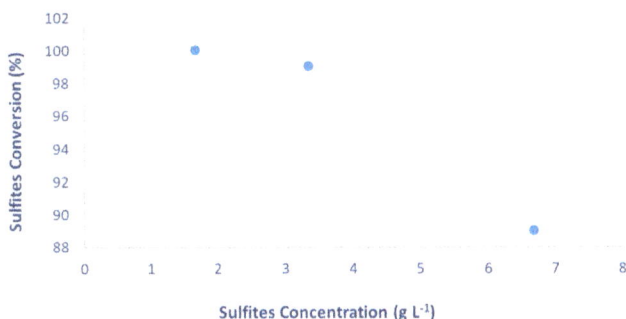

Figure 8. Sulphite conversion vs. their initial concentration after 24 h of reaction. P25 as a photocatalyst with 0.031 g·L^{-1} loading and at pH = 13/14.

2.2.4. Comparison between Different Photocatalysts

Flame spray pyrolysis allows the synthesis of titanium dioxide nanoparticles characterized by high surface area and high thermal stability [43–46]. Due to the abundance of oxygen and high temperature in the FSP reactor, the nanoparticles produced by FSP are typically fully oxidized and highly crystalline. The simple synthesis procedure permits the rapid and continuous production of the catalyst.

FSP titanium dioxide prepared in our lab has been tested for comparison with the commercial P25 titania. The samples were compared using 1.66 g L^{-1} of HS, with the selected catalyst loading (0.031 g L^{-1}) at 2 different pH conditions (Figure 9). Also in this case the conditions were selected to obtain significant amounts of products in both liquid and gas phase to check the effect of the other variables on both the mechanisms.

The results confirm also for the FSP titania a very limited productivity at neutral pH and a good productivity at basic pH. Slightly higher productivity of FSP has been partly attributed to its higher surface area (67 m^2 g^{-1} for FSP, 45 m^2 g^{-1} for P25), which increases the surface reactions rate, though being almost indifferent as for the main photochemical steps. Furthermore, the quite high activity of FSP and P25 catalysts has been attributed to the enhanced charge separation at the anatase-rutile interface which acts as charge traps (hence higher capacitance). This effect is much more remarkable for FSP at low catalyst loading, compared to the commercial TiO$_2$ catalysts. According to previous studies [43,46], short flame residence time in extreme conditions, arising from high temperatures, may produce a metastable phase and also a small concentration of defect states in the bandgap due to a Ti^{4+} stoichiometry deficiency, thus, enabling electron-hole pair generation as well as acting as photocharge trap defects [47]. The enhanced photocatalytic performance of FSP catalyst has been confirmed when varying catalyst concentration, which may result in smaller average primary particles and agglomerates and decreases light scattering (Figure 10) [45]. These results imply that the flame spray pyrolysis is a promising technique to produce catalyst that can be employed industrially for this application. TiO$_2$ P25 is also prepared through a flame synthesis, which however makes use of a different precursor and particle formation mechanism.

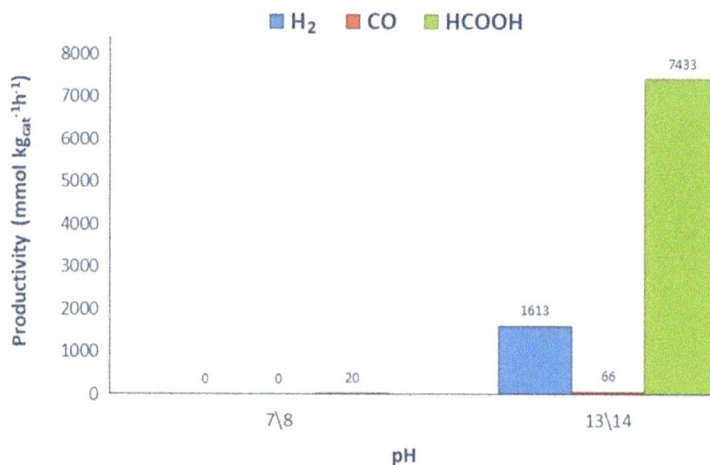

Figure 9. Influence of pH over productivity (0.031 g L^{-1} of FSP and 1.66 g L^{-1} HS).

Metals addition is a common strategy to improve visible light harvesting and to enhance the separation of photogenerated charges. We have investigated the performance of gold nanoparticles on the productivity and selectivity of products. 0.2 wt% Au loading was selected based on previous screening [37]. Au-P25 has been tested maintaining a fixed value of HS 1.66 g L^{-1}, basic conditions (pH 13/14) and, also in this case, variable catalyst amount. Figure 10 reports the overall comparison of productivity of all the tested photocatalysts.

The comparison of different catalysts maintaining the highest catalyst concentration in the reactor (0.5 g L^{-1}), demonstrates that adding gold to P25 increases the selectivity towards secondary products (hydrogen and CO) with respect to bare P25 and even FSP. The gold doped catalyst is characterized by higher visible light absorption, which positively affects the light harvesting ability and consequently the overall productivity. Furthermore, gold may act as electron trap to improve the charge separation efficiency. This increases the photocatalyst effectiveness for all the reaction steps depicted in the reaction schemes (vide supra).

CO can be either obtained by (i) direct photoreduction of CO_2, or (ii) as a product of photoreforming of the organic compounds obtained in liquid phase by CO_2 photoreduction, or even (iii) by catalytic reduction of CO_2 by using the photogenerated H_2. The productivity trend of CO and H_2 are so similar to suggests that both species are produced by photoreforming of the primary organic products of photoreduction accumulated in the liquid phase.

On the contrary, for 0.2 wt% Au-P25 decreasing the catalyst concentration results in decreasing selectivity towards hydrogen production with respect to P25, balanced by a significant increase of the productivity to HCOOH (Figure 10). The enhancement of productivity is due to the strong electric fields created by the surface plasmon resonance of the Au nanoparticles, which excite electron-hole pairs locally in the TiO_2 and produce a number of additional photocatalytic reaction products at a rate several orders of magnitude higher than the normal incident light [47]. In this wavelength range, both the excited electrons in Au and TiO_2 contribute to the reduction of CO_2 with H_2O [47].

Overall, by calculating the amount of energy stored as in Table 2, there is no appreciable difference between the use of the FSP catalyst and the 0.2 wt% Au/P25 one, both being more efficient than P25 from this point of view.

Finally, Table 3 gives a comprehensive comparison of the different TiO_2 based photocatalysts used for CO_2 photoreduction and their productivity and selectivity, in comparison with the present work. The comparison with the relevant literature reports confirm the validity of the presently adopted high pressure photoreduction apparatus, which is able to outperform most results by various orders of magnitude.

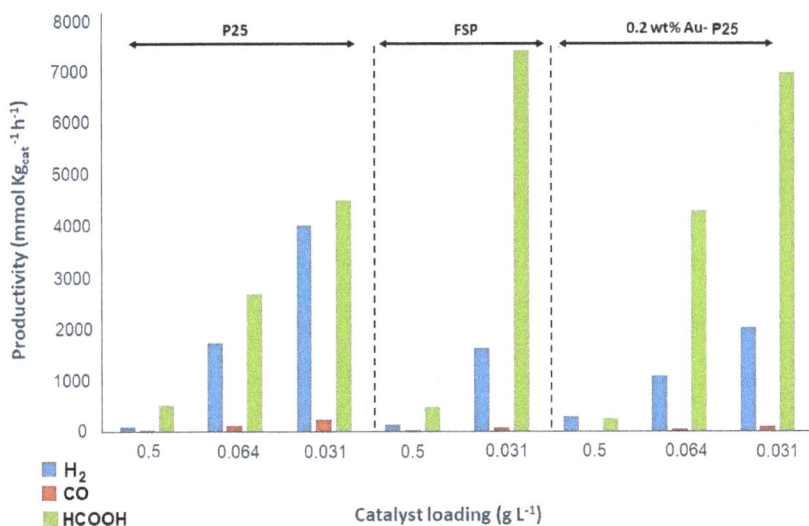

Figure 10. Products productivity with respect to different catalysts in different loadings with HS = 1.66 g L^{-1} and at pH = 13/14.

Table 3. Comparison of the photocatalytic performance for CO_2 photoreduction of TiO_2-based photocatalysts obtained with different techniques.

Strategy	Photocatalysts	Synthesis Method	Reaction Conditions	Activity	Ref
Increasing Surface area	Anatase TiO_2 with co-exposed (001) and (101) facets	Adjusting HF amounts in the solvothermal method	10 g L^{-1} catalysts; 300 W Xe arc lamp CO_2 and H_2O vapour were in-situ generated by the reaction of $NaHCO_3$ and HCl aqueous solution	The highest CH_4 generation rate was 1.35 µmol g^{-1} h^{-1}	[47]
Surface Defects	$Cu_{(I)}/TiO_{2-x}$ nanoparticles	Precipitation followed by thermal treatment	50 mg catalysts under 2 mL min^{-1} CO_2 flow; 150 W solar simulator (B90 mW cm^{-1})	$Cu_{(I)}/TiO_{2-x}$ exhibited the maximum CO production rate of 4.3 µmol g^{-1} h^{-1}	[48]
Surface basic sites	$NaOH–TiO_2$ composites	Impregnation method	Gas phase photoreduction with 80 mg catalysts in 80 kPa CO_2 in the presence of H_2O vapour; 300 W Xe lamp	Maximum CH_4 productivity 8.7 µmol g^{-1} h^{-1}	[49]
	$Pt–MgO/TiO_2$	Photo deposition and impregnation	20 mg catalyst on Teflon holder, in 2.0 MPa CO_2 with H_2O vapour; 100 W Xe lamp (λ = 320–780 nm)	Pt-1.0 wt% MgO/TiO_2 exhibited the highest CH_4 amount of 11 µmol g^{-1} h^{-1}	[50]
	Amine-functionalized TiO_2 by using monoethanolamine (MEA)	Solvothermal method	20 mg catalysts in 0.1 mL H_2O; Xe lamp	CO and CH_4 66.7 and 8.61 ppm h^{-1}, respectively	[51]
Surface noble-metal co-catalysts	3.0 wt% CuO/TiO_2	Impregnation and sonication	300 mg catalysts in 300 mL 1 M $KHCO_3$; CO_2 bubbled for 30 min to reach saturation; 10 W UV lamp	Methanol 442.5 µmol g^{-1} h^{-1}	[52]
	$Pd–TiO_2$	Photochemical deposition	150 mg catalysts in 1.5 mL H_2O; 500 W Hg lamp with a filter (λ > 310 nm)	$Pd–TiO_2$ exhibited a preferential generation of CH_4 instead of CO for bare TiO_2	[53]
	$Pt–TiO_2$ columnar films	Aerosol chemical vapour deposition	CO_2 and water vapour 3 mL min^{-1}; 400 W Xe lamp in the UV range (250–388 nm)	Selective formation of CH_4 as a main product with a yield of 1361 µmol g^{-1} h^{-1}	[54]
	0.2 wt% $Au–TiO_2$ P25	Impregnation precipitation	0.031 g L^{-1} catalyst with medium-pressure 125 W Hg vapour lamp with a range of emission $254 \leq \lambda \leq 364$ nm	HCOOH, CO and H_2 6980, 84 and 2018 µmol g^{-1} h^{-1}, respectively	This work
Semiconductor Systems	Rutile TiO_2 nanoparticle modified anatase TiO_2 nanorods (TiO_2-RMA)	Synthesis	Dispersion solution of catalyst and water with bubbling CO_2 until saturation point; 300 W Hg lamp	CH_4 2.36 µmol g^{-1} h^{-1}	[55]
	$CoPc-TiO_2$	Improved sol–gel method using a homogeneous hydrolysis technique	The suspension of catalyst powder in NaOH solution with CO_2 bubbling until saturation; 500 W tungsten–halogen lamp	HCOOH 28 µmol g^{-1} h^{-1} as a main product	[56]
	$Cu-TiO_2$	Sol–gel process	Suspension of catalyst powder in NaOH solution with CO_2 bubbling until saturation; Hg lamp (254 nm UVC or 365 nm UVA)	CH_3OH yield under UVC 600 µmol g$_{cat}^{-1}$. Under UVA 10 µmol g$_{cat}^{-1}$	[57]
	$Cu-TiO_2$	Sol–gel method using a homogeneous hydrolysis technique	Suspension of catalyst powder with 2.0 wt% in NaOH solution and CO_2 bubbling until saturation; 8 W Hg lamp with max emission at 254 nm	CH_3OH 19.6 µmol g^{-1} h^{-1}	[58]
	TiO_2 powder		0.8 g L^{-1} TiO_2 powder with CO_2 up to 9 MPa with Xe lamp 990 W, and the light intensity of 0.96 kW m^{-2} for 5 h	HCOOH 1.8 µmol g^{-1} h^{-1}	[59]
	$Rh-TiO_2$	Impregnation method	CO_2 (150 mmol) and H_2 (50 mmol) at 25 kPa. 500 W ultrahigh-pressure mercury lamp	CO/CH_4 5.2 µmol g^{-1} h^{-1}	[60]

Table 3. *Cont.*

Strategy	Photocatalysts	Synthesis Method	Reaction Conditions	Activity	Ref
	TiO$_2$-P25		TiO$_2$ powder suspended in iso-propyl alcohol solution as a hole scavenger and irradiated with a Xe lamp.	CH$_4$ 1.3 µmol g^{-1}	[61]
	TiO$_2$-P25		0.031 g L^{-1} catalyst with medium-pressure Hg vapour lamp with a range of emission 254 $\leq$ λ $\leq$ 364 nm	HCOOH and H$_2$ 4499 and 4000 µmol g^{-1} h^{-1}	This work
	TiO$_2$-FSP	Flame spray pyrolysis	0.031 g L^{-1} catalyst with medium-pressure Hg vapour lamp with a range of emission 254 $\leq$ λ $\leq$ 364 nm	HCOOH and H$_2$ 7433 and 1613 µmol g^{-1} h^{-1}	This work

3. Experimental

3.1. Materials Preparation

TiO$_2$ samples were prepared in dense nanoparticles form by FSP [44,62] and compared with a commercial P25 sample supplied by Evonik (code P25).

The FSP samples was prepared using a home-developed apparatus, composed of a burner which is co-fed with the titania precursor solution and 5 L/min of oxygen and the flame is ignited and sustained by a ring of flamelets (0.5 L/min CH$_4$ + 1 L/min of O$_2$). The solution of the oxide precursor in organic solvent is fed through a syringe pump at constant feeding rate of 2.5 mL/min in to the burner. The Titanium Isopropoxide (Sigma Aldrich, pur. 97%, St. Louis, MO, USA) as TiO$_2$ precursor was dissolved in o-xylene and Propionic acid (Sigma Aldrich, pur. 97%, St. Louis, MO, USA) with a 0.4 M concentration and injected through the burner. The pressure drop at the burner nozzle was 1.5 bar.

The gold doped TiO$_2$ samples (Au-P25) were prepared by a modified deposition-precipitation method using urea and a chemical reductant. 1 g of commercial TiO$_2$ (Degussa P25, 45 m^2 g^{-1}) was dispersed in distilled water (100 mL) then 5 g of urea (Aldrich, >99%, St. Louis, MO, USA). NaAuCl$_4$·2H$_2$O solution (Aldrich, 99.99%, St. Louis, MO, USA) was added to the suspension and left under vigorous stirring for 4 h at 80 °C. The catalyst was filtered and washed several times with distilled water. The collected sample after first washing was suspended in distilled water and a freshly prepared solution 0.1 M of NaBH$_4$ (Fluka, >96%, Bucharest, Romania) was added (NaBH$_4$/Au = 4 mol/mol) under vigorous stirring at room temperature. The sample was filtered, washed and dried at 100 °C for 4 h. Atomic Absorption Spectroscopy (AAS) analysis (Perkin Elmer 3100 instrument, Champaign, IL, USA) was performed to assess the final composition of Au-P25 catalysts: 0.2 wt% Au-P25, which proved the most active in a preliminary catalyst screening [63].

3.2. Materials Characterization

X-ray diffraction (XRD) analyses were performed by the Rigaku D III-MAX horizontal-scan powder diffractometer (Tokyo, Japan) using Cu-Kα radiation with a graphite monochromator on the diffracted beam.

N$_2$ adsorption and desorption isotherms of samples were collected with a Micromeritics ASAP2020 apparatus (Norcross, GA, USA).

Diffuse Reflectance (DR) UV-Vis spectra of samples were measured on a Cary 5000 UV-Vis-NIR spectrophotometer (Varian instruments, Santa Clara, CA, USA) in the range of 200–800 nm.

TPR analysis was carried out on a bench scale apparatus by flowing 40 mL/min of a 10 vol% H$_2$/N$_2$ mixture, while heating the sample by 10 °C/min up to 700 °C. The gas outflowing the quartz reactor was analysed with a TCD detector after entrapping the possibly formed water.

The TEM specimens were prepared by dispersing the catalyst powder on TEM grids coated with holey carbon film. They were examined in a FEI Titan 80–300 electron microscope equipped with CEOS image spherical aberration corrector, Fischione model 3000 high angle annular dark field (HAADF) scanning transmission electron microscopy (STEM) detector (Portland, OR, USA).

3.3. Photoreactor and Testing Conditions

All the experimental activity tests have been performed using an innovative pressurized batch photo-reactor which has been discussed in detail elsewhere [37,64,65]. The cylinder-shaped reactor made of AISI 316 stainless steel can operate up to 20 bar at temperatures up to 90 °C. The temperature is kept constant through a double-walled thermostatic system. The internal capacity of the reactor is ca. 1.3 L, filled with ca. 1.2 L solution. Continuous stirring inside the reactor is provided by a magnetic stirrer placed underneath up to 400 RPM to ensure the optimal dispersion of the catalyst in the liquid phase.

The radiation source is a medium-pressure 125 W Hg vapour lamp with a range of emission between 254 nm $\leq \lambda \leq$ 364 nm, with maximum emission at this latter wavelength. An air circulation system has been used to cool the lamp. The power of irradiation directly depends on the flow rate of the cooling pressurized air. Therefore, the best cooling condition for the optimum lamp lifetime with the maximum irradiation power has been selected. The emitted power was periodically measured by means of a photoradiometer (Delta OHM HD2102.2, Padua, Italy) and corresponds to ca. 104 W m^{-2} at the bottom of the source.

For the optimization of the best amount of catalyst, several concentrations have been chosen (ca. 0.5, 0.25, 0.125, 0.064, 0.031 g L^{-1}). The catalysts have been loaded with a suspension of bi-distilled water in the reactor. The best saturation condition has been settled overnight with the CO_2 saturation pressure of 7 bar and temperature of 80 °C, based on previous studies [26,37]. Testing was carried out under the same conditions, if not otherwise specified. Such a pressure and temperature allow to obtain a broad products spectrum both in gas and liquid phase, so they were set as optimal to investigate the effect of other parameters on productivity and selectivity to all the products.

Na_2SO_3 has been used as HS in different amounts (ca. 1.66, 3.34, 6.68 g L^{-1}) to understand its effect on productivity and selectivity to the various products. As expected, negligible productivity both in the liquid and in the gas phase has been observed without its addition. The photoreaction has been started by switching on the lamp for the 24 h of the reaction time.

Liquid products have been analysed by taking samples at the end of the reaction. For analysing the liquid products, HPLC (Agilent 1220 Infinity, with a column Alltech OA-10308, 300 mm_7.8 mm, Palo Alto, CA, USA), equipped with both UV and refractive index (Agilent 1260 Infinity, Palo Alto, CA, USA) detectors have been used. Aqueous H_3PO_4 solution (0.1 wt%) was used as the eluent. The gas products were collected in the headspace of the photoreactor and analysed by a gas chromatograph (Agilent 7890, Palo Alto, CA, USA) equipped with a TCD detector with the proper set up configuration for the quantification of H_2, CH_4 and polar/non-polar light gases.

4. Conclusions

The high pressure photoreduction of CO_2 in water has been studied under different operating conditions, investigating the role of catalyst concentration, varying the amount of hole scavenger and the effect of adding gold on productivity and selectivity. A comparison between different flame-based techniques for the preparation of TiO_2 was also done, that is, TiO_2 prepared by FSP and P25.

The hole scavenger plays a crucial role in the selective formation of gas products (CO and H_2) in the course of reaction time. In the presence of HS, photoreduction has been obtained in the liquid phase by formation of HCOOH as a main product. The consumption of the HS, instead, results in the consecutive photoreforming of the organic compounds accumulated in the liquid phase, with formation of secondary products, H_2 and CO, in the gas phase.

0.2 wt%-Au-P25 and TiO_2-FSP showed higher productivity for HCOOH with respect to TiO_2-P25. The method of synthesizing FSP nanoparticles may results in formation of metastable phase and defects which can further enhance the electron-hole pair generation and increasing the lifetime of photogenerated charges. Instead, the surface Plasmon resonance effect by doping Au on P25 can be considered as a main reason for higher HCOOH productivity in the presence of 0.2 wt%-Au-P25, with respect to bare TiO_2 P25.

Overall, appreciable amounts of energy per unit time have been stored through this reaction. The operating conditions should be tuned in order to drive the reaction towards the maximization of energy storage (high catalyst and HS concentrations) or the selection of the desired products.

Author Contributions: E.B. and A.T. Investigation, Data Curation, Writing-Original Draft Preparation; A.V. Investigation, Data Curation; C.P. Writing-Review & Editing; L.P. and G.R. Supervision, Project Administration, Funding Acquisition; I.R. Supervision, Project Administration, Funding Acquisition, Writing-Review & Editing.

Funding: This research was funded by Fondazione Cariplo and Regione Lombardia, grant number 2016-0858–"Up-Unconventional Photoreactors" and by MIUR through the PRIN2015 grant (20153T4REF).

Acknowledgments: The valuable help of the graduating student M.F. and of W.W. is gratefully acknowledged. We acknowledge Karlsruhe Nano Micro Facility for TEM.

Conflicts of Interest: The authors declare no conflict of interest.

References

1. Centi, G.; Perathoner, S. Opportunities and prospects in the chemical recycling of carbon dioxide to fuels. *Catal. Today* **2009**, *148*, 191–205. [CrossRef]

2. Jessop, G.P.; Ikariya, T.; Noyori, R. Homogeneous catalytic hydrogenation of supercritical carbon dioxide. *Nature* **1994**, *368*, 231–233. [CrossRef]

3. Olajire, A.A. Valorization of greenhouse carbon dioxide emissions into value-added products by catalytic processes. *J. CO_2 Util.* **2013**, *3–4*, 74–92. [CrossRef]

4. Wang, W.; Soulis, J.; Yang, Y.J.; Biswas, P. Comparison of CO_2 Photoreduction Systems: A Review. *Aerosol Air Qual. Res.* **2014**, *14*, 533–549. [CrossRef]

5. Yuan, L.; Xu, Y.-J. Photocatalytic conversion of CO_2 into value-added and renewable fuels. *Appl. Surf. Sci.* **2015**, *342*, 154–167. [CrossRef]

6. Linsebigler, A.L.; Lu, G.; Yates, J.T. Photocatalysis on TiO_2 Surfaces: Principles, Mechanisms, and Selected Results. *Chem. Rev.* **1995**, *95*, 735–758. [CrossRef]

7. Li, X.; Wen, J.; Low, J.; Fang, Y.; Yu, J. Design and fabrication of semiconductor photocatalyst for photocatalytic reduction of CO_2 to solar fuel. *Sci. China Mater.* **2014**, *57*, 70–100. [CrossRef]

8. Chang, X.; Wang, T.; Zhang, P.; Zhang, J.; Li, A.; Gong, J. Enhanced Surface Reaction Kinetics and Charge Separation of p–n Heterojunction $Co_3O_4/BiVO_4$ Photoanodes. *J. Am. Chem. Soc.* **2015**, *137*, 8356–8359. [CrossRef] [PubMed]

9. Zhai, Q.; Xie, S.; Fan, W.; Zhang, Q.; Wang, Y.; Deng, W.; Wang, Y. Photocatalytic Conversion of Carbon Dioxide with Water into Methane: Platinum and Copper(I) Oxide Co-catalysts with a Core-Shell Structure. *Angew. Chem.* **2013**, *52*, 5776–5779. [CrossRef] [PubMed]

10. White, J.L.; Baruch, M.F.; Pander, J.E.; Hu, Y.; Fortmeyer, I.C.; Park, J.E.; Zhang, T.; Liao, K.; Gu, J.; Yan, Y.; et al. Light-Driven Heterogeneous Reduction of Carbon Dioxide: Photocatalysts and Photoelectrodes. *Chem. Rev.* **2015**, *115*, 12888–12935. [CrossRef] [PubMed]

11. Habisreutinger, S.N.; Schmidt-Mende, L.; Stolarczyk, J.K. Photocatalytic Reduction of CO_2 on TiO_2 and Other Semiconductors. *Angew. Chem. Int. Ed.* **2013**, *52*, 7372–7408. [CrossRef] [PubMed]

12. Xie, S.; Zhang, Q.; Liu, G.; Wang, Y. Photocatalytic and photoelectrocatalytic reduction of CO_2 using heterogeneous catalysts with controlled nanostructures. *Chem. Commun.* **2015**, *52*, 35–59. [CrossRef] [PubMed]

13. Ma, Y.; Wang, X.; Jia, Y.; Chen, X.; Han, H.; Li, C. Titanium Dioxide-Based Nanomaterials for Photocatalytic Fuel Generations. *Chem. Rev.* **2014**, *114*, 9987–10043. [CrossRef] [PubMed]

14. Indrakanti, V.P.; Kubicki, J.D.; Schobert, H.H. Photoinduced activation of CO_2 on Ti-based heterogeneous catalysts: Current state, chemical physics-based insights and outlook. *Energy Environ. Sci.* **2009**, *2*, 745. [CrossRef]

15. Gattrell, M.; Gupta, N.; Co, A. A review of the aqueous electrochemical reduction of CO_2 to hydrocarbons at copper. *J. Electroanal. Chem.* **2006**, *594*, 1–19. [CrossRef]

16. Michalkiewicz, B.; Majewska, J.; Ka, G.; Bubacz, K.; Mozia, S.; Morawski, A.W. Reduction of CO_2 by adsorption and reaction on surface of TiO_2-nitrogen modified photocatalyst. *J. CO_2 Util.* **2014**, *5*, 47–52. [CrossRef]

17. Izumi, Y. Recent advances (2012–2015) in the photocatalytic conversion of carbon dioxide to fuels using solar energy: Feasibilty for a new energy. *ACS Symp. Ser.* **2015**, *1194*, 1–46.

18. He, H.; Zapol, P.; Curtiss, L.A. A Theoretical Study of CO_2 Anions on Anatase (101) Surface. *J. Phys. Chem. C* **2010**, *114*, 21474–21481. [CrossRef]

19. He, H.; Zapol, P.; Curtiss, L.A. Computational screening of dopants for photocatalytic two-electron reduction of CO_2 on anatase (101) surfaces. *Energy Environ. Sci.* **2012**, *5*, 6196–6205. [CrossRef]

20. Chen, B.-R.; Nguyen, V.-H.; Wu, J.C.S.; Martin, R.; Koči, K. Production of renewable fuels by the photohydrogenation of CO_2: Effect of the Cu species loaded onto TiO_2 photocatalysts. *Phys. Chem. Chem. Phys.* **2016**, *18*, 4942–4951. [CrossRef] [PubMed]

21. Koci, K.; Matteju, K.; Obalovà, L.; Krejcikovà, S.; Lacny, Z.; Plachà, D.; Capek, L.; Hospodkovà, A.; Solcovà, O. Effect of silver doping on the TiO_2 for photocatalytic reduction of CO_2. *Appl. Catal. B Environ.* **2010**, *96*, 239–244. [CrossRef]

22. Li, K.; An, X.; Hyeon, K.; Khraisheh, M.; Tang, J. A critical review of CO_2 photoconversion: Catalysts and reactors. *Catal. Today* **2014**, *224*, 3–12. [CrossRef]

23. Chen, X.; Shen, S.; Guo, L.; Mao, S.S. Semiconductor-based Photocatalytic Hydrogen Generation. *Chem. Rev.* **2010**, *110*, 6503–6570. [CrossRef] [PubMed]

24. Rossetti, I.; Villa, A.; Compagnoni, M.; Prati, L.; Ramis, G.; Pirola, C.; Bianchi, C.L.; Wang, W.; Wang, D. CO_2 photoconversion to fuels under high pressure: Effect of TiO_2 phase and of unconventional reaction conditions. *Catal. Sci. Technol.* **2015**, *5*, 4481–4487. [CrossRef]

25. Rossetti, I.; Villa, A.; Pirola, C.; Prati, L.; Ramis, G. A novel high-pressure photoreactor for CO_2 photoconversion to fuels. *RSC Adv.* **2014**, *4*, 28883–28885. [CrossRef]

26. Olivo, A.; Ghedini, E.; Signoretto, M.; Compagnoni, M.; Rossetti, I. Liquid vs. Gas Phase CO_2 Photoreduction Process: Which Is the Effect of the Reaction Medium? *Energies* **2017**, *10*, 1394. [CrossRef]

27. Rossetti, I.; Bahadori, E.; Tripodi, A.; Villa, A.; Prati, L.; Ramis, G. Conceptual design and feasibility assessment of photoreactors for solar energy storage. *Sol. Energy* **2018**. [CrossRef]

28. Spurr, R.A.; Myers, H. Quantitative Analysis of Anatase-Rutile Mixtures with an X-ray Diffractometer. *Anal. Chem.* **1957**, *29*, 760–762. [CrossRef]

29. Chauhan, R.; Kumar, A.; Chaudhary, R.P. Structural and optical characterization of Ag-doped TiO_2 nanoparticles prepared by a sol–gel method. *Res. Chem. Intermed.* **2012**, *38*, 1443–1453. [CrossRef]

30. Thommes, M.; Kaneko, K.; Neimark, A.V.; Olivier, J.P.; Rodriguez-reinoso, F.; Rouquerol, J.; Sing, K.S.W. Physisorption of gases, with special reference to the evaluation of surface area and pore size distribution (IUPAC Technical Report). *Pure Appl. Chem.* **2015**, *87*, 1051–1069. [CrossRef]

31. Seifvand, N.; Kowsari, E. Synthesis of Mesoporous Pd-Doped TiO_2 Templated by a Magnetic Recyclable Ionic Liquid for Efficient Photocatalytic Air Treatment. *Ind. Eng. Chem. Res.* **2016**, *55*, 10533–10543. [CrossRef]

32. Tan, S.T.; Chen, B.J.; Sun, X.W. Blueshift of optical band gap in ZnO thin films grown by metal-organic chemical-vapour deposition. *J. Appl. Phys.* **2005**, *98*, 013505. [CrossRef]

33. Li, P.; Hu, H.; Xu, J.; Jing, H.; Peng, H.; Lu, J. New insights into the photo-enhanced electrocatalytic reduction of carbon dioxide on MoS_2-rods/TiO_2 NTs with unmatched energy band. *Appl. Catal. B Environ.* **2014**, *147*, 912–919. [CrossRef]

34. Mogal, S.I.; Gandhi, V.G.; Mishra, M.; Tripathi, S.; Shripathi, T.; Joshi, P.; Shah, D.O. Single-Step Synthesis of Silver-Doped Titanium Dioxide: Influence of Silver on Structural, Textural, and Photocatalytic Properties. *Ind. Eng. Chem. Res.* **2014**, *53*, 5749–5758. [CrossRef]

35. Izumi, Y. Recent advances in the photocatalytic conversion of carbon dioxide to fuels with water and/or hydrogen using solar energy and beyond. *Coord. Chem. Rev.* **2013**, *257*, 171–186. [CrossRef]

36. Ao, C.H.; Lee, S.C.; Yu, J.Z.; Xu, J.H. Photodegradation of formaldehyde by photocatalyst TiO_2: Effects on the presences of NO, SO_2 and VOCs. *Appl. Catal. B Environ.* **2004**, *54*, 41–50. [CrossRef]

37. Galli, I.R.F.; Compagnoni, M.; Vitali, D.; Pirola, C.; Bianchi, C.L.; Villa, A.; Prati, L. CO_2 photoreduction at high pressure to both gas and liquid products over titanium dioxide. *Appl. Catal. B Environ.* **2017**, *200*, 386–391. [CrossRef]

38. Shen, J.; Kortlever, R.; Kas, R.; Birdja, Y.Y.; Diaz-morales, O.; Kwon, Y.; Ledezma-yanez, I.; Schouten, K.J.P.; Mul, G.; Koper, M.T.M. Electrocatalytic reduction of carbon dioxide to carbon monoxide and methane at an immobilized cobalt. *Nat. Commun.* **2015**, *6*, 1–8. [CrossRef] [PubMed]

39. Weng, Z.; Jiang, J.; Wu, Y.; Wu, Z.; Guo, X.; Materna, K.L.; Liu, W.; Batista, V.S.; Brudvig, G.W.; Wang, H. Electrochemical CO_2 Reduction to Hydrocarbons on a Heterogeneous Molecular Cu Catalyst in Aqueous Solution. *J. Am. Chem. Soc.* **2016**, *138*, 8076–8079. [CrossRef] [PubMed]

40. Manthiram, K.; Beberwyck, B.J.; Alivisatos, A.P. Enhanced Electrochemical Methanation of Carbon Dioxide with a Dispersible Nanoscale Copper Catalyst. *J. Am. Chem. Soc.* **2014**, *136*, 13319–13325. [CrossRef] [PubMed]

41. Xie, M.S.; Xia, B.Y.; Li, Y.; Yan, Y.; Yang, Y.; Sun, Q.; Chan, S.H.; Fishere, A.; Wang, X. Amino acid modified copper electrodes for the enhanced selective electroreduction of carbon dioxide towards hydrocarbons. *Energy Environ. Sci.* **2016**, *9*, 1687–1695. [CrossRef]

42. Kočí, K.; Obalová, L.; Matějová, L.; Plachá, D.; Lacný, Z.; Jirkovský, J.; Šolcová, O. Effect of TiO_2 particle size on the photocatalytic reduction of CO_2. *Appl. Catal. B Environ.* **2009**, *89*, 494–502. [CrossRef]

43. Teoh, W.Y. A Perspective on the Flame Spray Synthesis of Photocatalyst Nanoparticles. *Materials* **2013**, *6*, 3194–3212. [CrossRef] [PubMed]

44. Compagnoni, M.; Lasso, J.; di Michele, A.I.; Rossetti, I. Flame-pyrolysis-prepared catalysts for the steam reforming of ethanol. *Catal. Sci. Technol.* **2016**, *6*, 6247–6256. [CrossRef]

45. Giacomo, L.; Maria, B.; Diamanti, V.; Sansotera, M.; Pia, M.; Walter, P.; Paolo, N. Immobilized TiO_2 nanoparticles produced by flame spray for photocatalytic water remediation. *J. Nanopart. Res.* **2016**, *18*, 238.

46. Bettini, L.; Dozzi, M.; della Foglia, F.; Chiarello, G.; Selli, E.; Lenardi, C.; Piseri, P.; Milani, P. Mixed-phase nanocrystalline TiO_2 photocatalysts produced by flame spray pyrolysis. *Appl. Catal. B Environ.* **2015**, *178*, 226. [CrossRef]

47. Yu, J.; Low, J.; Xiao, W.; Zhou, P.; Jaroniec, M. Enhanced Photocatalytic CO_2 Reduction Activity of Anatase TiO_2 by Coexposed {001} and {101} Facets. *J. Am. Chem. Soc.* **2014**, *136*, 8839–8842. [CrossRef] [PubMed]

48. Liu, L.; Zhao, C.; Li, Y. Spontaneous Dissociation of CO_2 to CO on Defective Surface of $Cu(I)/TiO_{2-x}$ Nanoparticles at Room Temperature. *J. Phys. Chem. C* **2012**, *116*, 7904–7912. [CrossRef]

49. Meng, X.; Ouyang, S.; Kako, T.; Li, P.; Yu, Q.; Wang, T.; Ye, J. TiO_2 without loading noble metal cocatalyst. *Chem. Commun.* **2014**, *50*, 11517–11519. [CrossRef] [PubMed]

50. Xie, S.; Wang, Y.; Zhang, Q.; Fan, W.; Deng, W.; Wang, Y. Photocatalytic reduction of CO_2 with H_2O: Significant enhancement of the activity of Pt–TiO_2 in CH_4 formation by addition of MgO. *Chem. Commun.* **2013**, *49*, 2451–2453. [CrossRef] [PubMed]

51. Liao, Y.; Cao, S.; Yuan, Y.; Gu, Q.; Zhang, Z.; Xue, C. Efficient CO_2 Capture and Photoreduction by Amine-Functionalized TiO_2. *Chem. Eur. J.* **2014**, *20*, 10220–10222. [CrossRef] [PubMed]

52. Nasution, H.W.; Purnama, E.; Kosela, S.; Gunlazuardi, J. Photocatalytic reduction of CO_2 on copper-doped Titania catalysts prepared by improved-impregnation method. *Catal. Commun.* **2005**, *6*, 313–319. [CrossRef]

53. Ishitani, O.; Inoue, C.; Suzuki, Y.; Ibusuki, T. Photocatalytic reduction of carbon dioxide to methane and acetic acid by an aqueous suspension of metal-deposited TiO_2. *J. Photochem. Photobiol. A Chem.* **1993**, *72*, 269–271. [CrossRef]

54. Wang, W.; An, W.; Ramalingam, B.; Mukherjee, S.; Niedzwiedzki, D.M.; Gangopadhyay, S.; Biswas, P. Size and Structure Matter: Enhanced CO_2 Photoreduction Efficiency by Size-Resolved Ultrafine Pt Nanoparticles on TiO_2 Single Crystals. *J. Am. Chem. Soc.* **2012**, *134*, 11276–11281. [CrossRef] [PubMed]

55. Wang, P.Q.; Bai, Y.; Liu, J.Y.; Fan, Z.; Hu, Y.Q. One-pot synthesis of rutile TiO_2 nanoparticle modified anatase TiO_2 nanorods toward enhanced photocatalytic reduction of CO_2 into hydrocarbon fuels. *Catal. Commun.* **2012**, *29*, 185–188. [CrossRef]

56. Liu, S.; Zhao, Z.; Wang, Z. Photocatalytic reduction of carbon dioxide using sol–gel derived titania-supported CoPc catalysts. *Photochem. Photobiol. Sci.* **2007**, *6*, 695–700. [CrossRef] [PubMed]

57. Tseng, I.-H.; Wu, J.C.S.; Chou, H.-Y. Effects of sol–gel procedures on the photocatalysis of Cu/TiO_2 in CO_2 photoreduction. *J. Catal.* **2004**, *221*, 432–440. [CrossRef]

58. Tseng, I.-H.; Chang, W.-C.; Wu, J.C.S. Photoreduction of CO_2 using sol–gel derived titania and titania-supported copper catalysts. *Environ. Appl. Catal. B* **2002**, *37*, 37–48. [CrossRef]

59. Kaneco, S.; Kurimoto, H.; Shimizu, Y.; Ohta, K. Photocatalytic reduction of CO_2 using TiO_2 powders in supercritical fluid CO_2. *Energy* **1999**, *24*, 21–30. [CrossRef]

60. Kohno, Y.; Hayashi, H.; Takenaka, S.; Tanaka, T.; Funabiki, T.; Yoshida, S. Photo-enhanced reduction of carbon dioxide with hydrogen over Rh/TiO_2. *J. Photochem. Photobiol. A Chem.* **1999**, *126*, 117–123. [CrossRef]

61. Kaneco, S.; Shimizu, Y.; Ohta, K.; Mizuno, T. Photocatalytic reduction of high pressure carbon dioxide using TiO_2 powders with a positive hole scavenger. *J. Photochem. Photobiol. A Chem.* **1998**, *115*, 223–226. [CrossRef]

62. Chiarello, G.L.; Rossetti, I.; Forni, L. Flame-spray pyrolysis preparation of perovskites for methane catalytic combustion. *J. Catal.* **2005**, *236*, 251–261. [CrossRef]

63. Compagnoni, M.; Kondrat, S.A.; Chan-Thaw, C.E.E.; Morgan, D.J.; Wang, D.; Prati, L.; Dimitratos, N.; Rossetti, I. Spectroscopic Investigation of Titania Supported Gold Nanoparticles Prepared by a Modified DP Method for the Oxidation of CO. *ChemCatChem* **2016**, *8*, 2136–2145. [CrossRef]

64. Thamaphat, K.; Limsuwan, P.; Ngotawornchai, B. Phase Characterization of TiO_2 Powder by XRD and TEM. *Conf. Proc.* **2008**, *361*, 357–361.

65. Compagnoni, M.; Ramis, G.; Freyria, F.S.; Armandi, M.; Bonelli, B.; Rossetti, I. Innovative photoreactors for unconventional photocatalytic processes: The photoreduction of CO_2 and the photo-oxidation of ammonia. *Rend. Lincei* **2017**, *28*, 151–158. [CrossRef]

Article

Effect of In-Situ Dehydration on Activity and Stability of Cu–Ni–K₂O/Diatomite as Catalyst for Direct Synthesis of Dimethyl Carbonate

Dongmei Han [1], Yong Chen [2], Shuanjin Wang [2], Min Xiao [2,*], Yixin Lu [3] and Yuezhong Meng [1,2,*]

[1] School of Chemical Engineering and Technology, Sun Yat-sen University, Guangzhou 510275, China; handongm@mail.sysu.edu.cn

[2] The Key Laboratory of Low-Carbon Chemistry & Energy Conservation of Guangdong Province/State Key Laboratory of Optoelectronic Materials and Technologies, Sun Yat-sen University, Guangzhou 510275, China; chenyong@lesso.com (Y.C.); wangshj@mail.sysu.edu.cn (S.W.)

[3] Department of Chemistry & Medicinal Chemistry Program, Office of Life Sciences, National University of Singapore, Singapore 117543, Singapore; chmlyx@nus.edu.sg

* Correspondence: stsxm@mail.sysu.edu.cn (M.X.); mengyzh@mail.sysu.edu.cn (Y.M.); Tel./Fax: +86-20-8411-4113 (Y.M.)

Received: 31 July 2018; Accepted: 18 August 2018; Published: 23 August 2018

Abstract: An in-situ dehydrating system built in a continuous flow fixed-bed bubbling reactor for direct synthesis of dimethyl carbonate (DMC) was designed. 3A molecular sieve (MS) was selected as the ideal dehydrating agent and the water trapping efficiency was studied. The effect of dehydrating agent/catalyst ratio, the dehydrating temperature and pressure, as well as the space velocity on the direct DMC synthesis catalyzed by K₂O-promoted Cu–Ni was further investigated. These results demonstrated that 3A MS could effectively dehydrate the reaction system at the optimal conditions of 120 °C and 1.0 MPa with gas space velocity (GHSV) of 600 h⁻¹, thereby greatly shifting the reaction equilibrium toward high DMC yield. Higher DMC yield of 13% was achieved compared with undehydrated reaction. Moreover, the catalyst can be highly stabilized by 3A MS dehydration with stable performs over 22 h.

Keywords: alkali promoter; dimethyl carbonate; catalysis; carbon dioxide; dehydration

---

1. Introduction

The environment-friendly dimethyl carbonate has aroused great interest in fuel additives, polar solvents, and methylating and carbonylating agents [1–3]. It has been produced worldwide by several commercial methods such as ester exchange process [4,5], methanolysis of phosgene [6], and gas-phase oxidative carbonylation of methanol [7]. Recently, some green and economic dimethyl carbonate (DMC) synthesis routes have been studied all over the world. In these routes, DMC was directly synthesized from CH₃OH and CO₂ instead of using toxic, corrosive, flammable, and explosive gases such as phosgene, hydrogen chloride, and carbon monoxide as feedstock [8,9].

Over the last decades, improving the yield of DMC from the direct synthetic route has been mainly focused on catalyst development and optimization of reaction conditions. A large number of examples have been devoted to the direct synthesis of DMC from CO₂ and CH₃OH using organometallic compounds [10], CeO₂ [11], CeO₂-ZrO₂, Ce₀.₅Zr₀.₅O₂, Co₁.₅PW₁₂O₄₀ [12–14], or H₃PO₄–V₂O₅ catalyst [15], modified Cu–Ni bimetallic catalyst [8,16–19], ionic liquid [20], etc. The direct synthetic route of DMC was represented as follows:

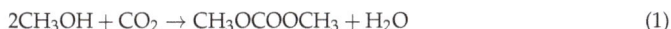

$$2CH_3OH + CO_2 \rightarrow CH_3OCOOCH_3 + H_2O \tag{1}$$

The drawback of this strategy is the low methanol conversion, which is ascribed to the thermodynamic limitations and/or catalyst deactivation by the water byproduct that hydrolyze the formed DMC, the high bond energy of CO_2, the reversible nature of the reaction, and the inability to utilize physical absorption agents of water such as zeolites, $CaCl_2$, and molecular sieves due to high operating temperatures and pressures, but no report had addressed these problems. Since the reaction is nonspontaneous (Equation (1)) and the reaction equilibrium is quickly established, the dehydrating agent can enhance the yield of DMC by shifting the equilibrium toward higher DMC yields.

For batch reaction, the dehydrating additives such as trimethyl orthoacetate [21], 2,2-dimethoxypropane [22], acetonitrile [23], butylene oxide [24], and a recyclable dehydrating tube packed with molecular sieves 3A [25] can improve the yield of DMC by shifting the equilibrium. However, the economy of this process, the activity of the catalyst, the separation of product and by-product become the new subjects for further research. For continuous flow reaction, these obstacles were evidently cleared away. Herein, a concept of in-situ water removal via the addition of an inorganic dehydrating agent during the direct DMC synthesis from methanol and CO_2 catalyzed by K_2O-promoted Cu–Ni is presented. The 3A MS is selected as the optimal dehydrating agent effects of the mass ratio of dehydrating agent and catalyst, the dehydrating temperature and pressure, together with the space velocity on the catalytic performance for direct synthesis of DMC were investigated and discussed.

2. Results and Discussion

2.1. Selection of the Dehydrating Agent

There have been many previous attempts to remove the water from the reaction of direct synthesis of DMC using $MgSO_4$, Na_2SO_4, $CaCl_2$, etc. as inorganic dehydrating agents, however, no successful result was obtained because of high reaction temperatures and pressures. In this paper, low-cost and readily available molecular sieve of 3A, 4A, and 5A are chosen as the dehydrating agents. Prior to use, the three molecular sieves were treated at 500 °C for 6 h, and then allowed to fully hydrate under saturation pressure of water vapor (25 °C), followed by dehydration from 50 to 600 °C. The weight loss of the samples was recorded by Thermogravimetric analyses (TGA). As shown in Figure 1, 3A molecular sieve shows 11 wt. % of water escaped out at the reaction temperature of 120 °C, indicating that the 3A molecular sieve is the best candidate in this issue.

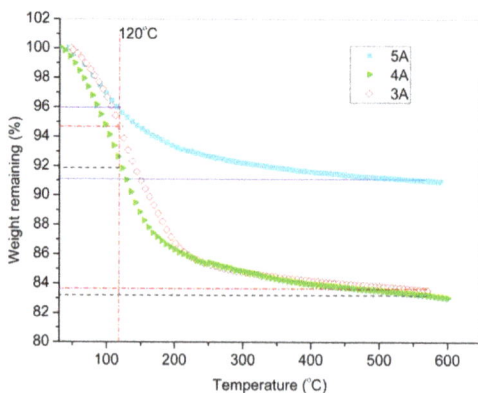

Figure 1. Thermogravimetric analyses (TGA) traces of molecular sieves (3A, 4A, and 5A) saturated by water vapor (1 atm, 25 °C).

2.2. Characterization of the Catalyst

In our previous work [26], it is investigated that the incorporation of alkali is conducive to the preparation of the catalysts precursor by decreasing the decomposition and reduction temperatures, which is favorable for the formation of a Nano-scale dispersion of bimetallic particles on the surface of supports. The well-dispersed characteristic in turn endows the catalyst with more lattice drawbacks and a polarized Cu-Ni lattice. It is proved that alkali doping can significantly improve the catalytic efficiency of Cu-Ni composites. Based on this, Cu-Ni-K_2O/diatomite catalysts are employed in this work. Temperature-programmed reduction and desorption (TPR, NH_3–TPD, CO_2–TPD) of the sample are all included in Figure 2. It can be seen that the catalyst precursor can be fully reduced below 450 °C with two closely combined peaks attributed to the reduction of CuO–NiO–K_2O solid composite. The NH_3–TPD and CO_2–TPD of the catalyst both exhibit one desorption peak around 200 °C, indicating that the catalyst has only one type of medium acid center and basic center, which are essential for direct catalytic synthesis of DMC. Figure 3 presented the powder X-ray diffraction of the samples. The reduced catalyst clearly shows four typical diffraction peaks of Cu, Ni or Cu–Ni alloy (2θ = 43.62 (111), 51.06 (200), 74.94 (220), and 91.04 (311)). Compared with 15%(2Cu–Ni)/diatomite (i.e., 15 CN/diatomite), the characteristic diffraction peaks of 15CN2K/diatomite became obviously broader than the undoped catalyst, which implies that the K-doped catalyst has better dispersion and smaller particle size than 15%(2Cu–Ni)/diatomite catalyst. According to Scherrer equation (D = Kλ/βcosθ, λ= 0.15406, and K = 0.89), the partical size of 15%(2Cu-Ni)/diatomite is about 22 nm and which of 15CN2K/diatomite is about 30 nm. Figure 4 displayed the SEM (a) and TEM (b) observation results of 15CN2K/diatomite. It shows that the Cu–Ni–K_2O composite homogenously covered the support and the single particle size of the catalyst is less than 50 nm.

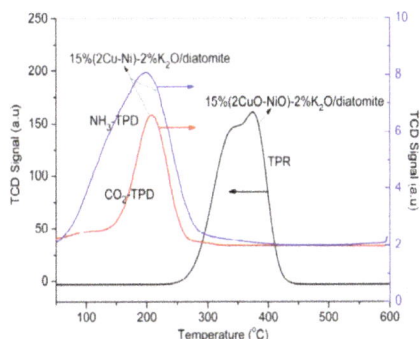

Figure 2. TPR, CO_2–TPD and NH_3–TPD of 15CN2K/diatomite.

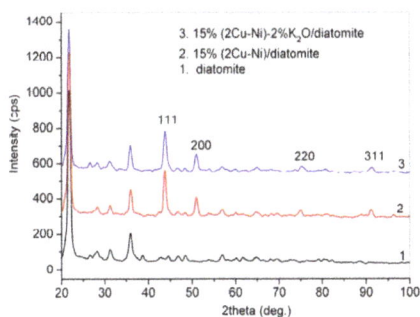

Figure 3. Powder X-ray diffraction patterns of the samples.

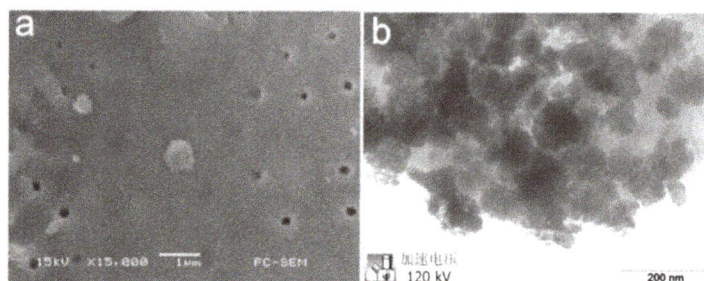

Figure 4. SEM (**a**) and TEM (**b**) of 15CN2K/diatomite catalyst.

2.3. Effect of Dehydration on Properties of the Catalyst

2.3.1. Effect of Mass Ratio of 3A MS and 15CN2K/Diatomite on the Activity of the Catalyst

As shown in Figure 5, the mass ratio of 3A MS and 15CN2K/diatomite ranges from 0 to 5. The average value of 4-h methanol conversion climbs from 7.55 to 8.41% and the corresponding DMC selectivity fluctuates from 90.3 to 89.9%. The optimal methanol conversion of 8.27% with highest DMC selectivity of 91.2% has been achieved with 3/1 of 3A MS and the catalyst. The results indicated that 3A MS can effectively dehydrate the reaction system and shift the reaction equilibrium toward high DMC yield. The mass ratio of 3/1 is most preferable for this reaction from the economic point of view.

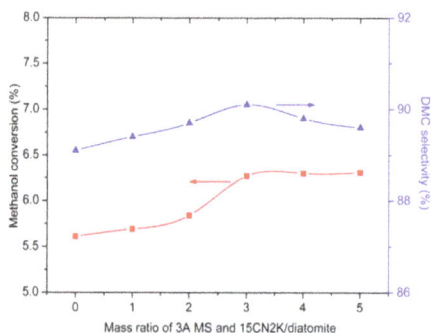

Figure 5. The effect of mass ratio of 3A MS and 15CN2K/diatomite on performance of 15CN2K/diatomite.

2.3.2. Effect of Dehydrating Temperature and Pressure on Properties of the Catalyst

The effect of dehydrating temperature on DMC synthesis was exhibited in Figure 6, the methanol conversion of this reaction is enhanced with the increase of temperature under the set pressure and space velocity. Moreover, the in-situ dehydrated catalyst performs superiorly to the catalyst without dehydration in methanol conversion. Nevertheless, DMC selectivity dropped dramatically over 140 °C due to the intensified side reaction of dimethyl ether (DME) and formic acid formation. Furthermore, the DMC selectivity of dehydrated catalyst collapsed more quickly at high temperature because 3A MS gradually fails to dehydrate the catalyst at high temperature, and 3A MS may serve as the catalyst for DME production. So 120 °C is chose in the reaction.

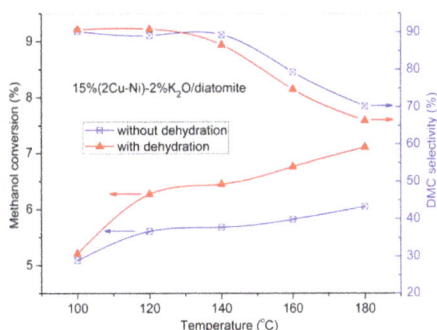

Figure 6. The effect of temperature on performance of catalyst with or without dehydration.

The effect of dehydrating pressure on DMC synthesis is presented in Figure 7. The methanol conversion is obviously improved with the increase of reaction pressure at fixed temperature and space velocity, and it levels off over 1.0 MPa. Compared with the undehydrated catalyst, the in-situ dehydrated catalyst shows higher activity, suggesting that the increased pressure is more favorable for dehydrating this reaction. The DMC selectivity keeps closely around 90% for undehydrated catalyst. For the dehydrated catalyst, it shows a little increase under 0.8 MPa and leveling off under 1.0 MPa, which implies that in-situ dehydration can produce more active sites for DMC catalytic synthesis. In general, the optimum conditions for effective dehydrating this reaction system are 120–140 °C and 1.0 MPa from the economic angle. Herein, 1.0 MPa is used in the following reaction.

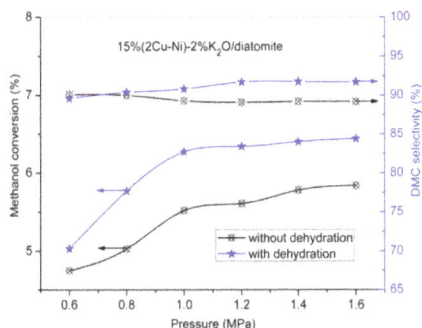

Figure 7. The effect of pressure on performance of catalyst with or without dehydration.

2.3.3. Effect of Space Velocity on Dehydrating the Catalyst

At fixed 120 °C and under 1.0 MPa, the methanol conversion of undehydrated catalyst is evidently lower than dehydrated catalyst. The methanol conversion of dehydrated catalyst decreases from 6.25 to 4.51% with increasing gas space velocity (GHSV) from 300 to 1500 h^{-1} (Figure 8). However, the methanol conversion of dehydrated catalyst goes up a little at first with increasing GHSV from 300 to 600 h^{-1}, and then it decreases gradually from 8.35 to 6.19% with increasing GHSV from 600 to 1500 h^{-1}. As for the two catalytic reaction processes, space velocity seems no effect on DMC selectivity which keeps around 90%. So 600 h^{-1} is chose in the reaction.

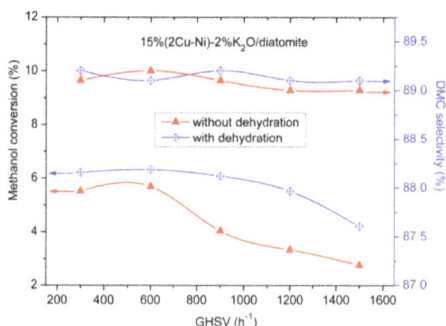

Figure 8. The effect of gas space velocity (GHSV) on performance of catalyst with or without dehydration.

To summarize, the optimal condition is that the mass ratio of 3/1 (3A MS and 15CN2K/diatomite), temperature of 120 °C, dehydrating pressure of 1.0 MPa with GHSV of 600 h^{-1}. In this condition, the DMC selectivity is 89.2%, and methanol conversion of 6.49%. Regarding the pathway of the catalysis, which is similar to our previous work [17,19]. In general, there are three types of active centers: Cu-Ni metal sites, Lewis acid sites and Lewis base sites. Firstly, Horizontal adsorption state of CO_2 can be formed under the synergistic action of Lewis acid sites and metal sites and this adsorption state is reactive. The addition of K_2O additive is more conducive to the adsorption of CO_2 on the catalyst surface [26]. Secondly, dissociated adsorption states of CH_3OH could be formed in the association of Lewis acid sites and Lewis base sites. After that, adsorption state of CO_2 reacts with dissociated adsorption states of CH_3OH to form DMC. The main product of CO_2 and CH_3OH on the surface of catalyst is DMC.

2.3.4. Effect of In-Situ Dehydration on Stability of the Catalyst

The effect of in-situ dehydration on stability of the catalyst is evaluated within 22 h at 120 °C and under 1.0 MPa with GHSV of 600 h^{-1} (Figure 9). The methanol conversion of the reaction without dehydration increases to 7.55% at the beginning, and then gradually decreases to 6.18% within 8 h, after that it sharply collapses and deactivates at the end of the evaluation. The methanol conversion of reaction with dehydrating process keeps around 8% within 10 h, followed by decreasing to 1.84% at the end. The DMC selectivity of the reaction without dehydration decreases slowly from 88% to 83%, thereafter rapidly falls down to about 71%. However, the DMC selectivity with dehydration maintains over 88% within 13 h; and finally decreases to 78% at the end of the evaluation. By comparison, it is apparent that the catalyst dehydrated by 3A MS exhibited much higher activity and longer stability than the catalyst without dehydration, which indicated 3A MS can effectively dehydrate this catalytic reaction system at lower temperature and pressure. The main reason for the deactivation of the catalyst is that the chemical environment of the active species on the surface of the catalyst changes after a period of catalytic reaction, which means that the active site of the catalyst is gradually deactivated during the reaction, resulting in a decrease in the yield of DMC. The reasons for the deactivation of catalysts are believed to result from the reaction between the catalyst with formed water, followed by the oxidation of catalyst.

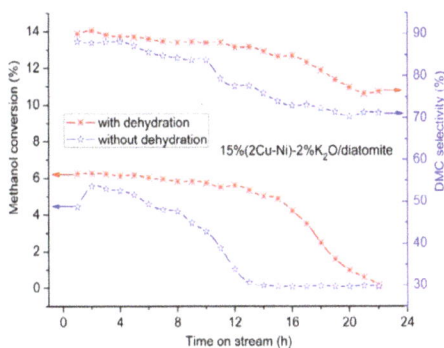

Figure 9. Stability study of the catalyst evaluated with or without dehydration.

3. Experimental

3.1. Catalyst Preparation

Natural diatomite was pretreated by calcining at 500 °C for 3 h, soaking in 5% hydrochloric acid for 24 h, washing by deionized water and drying overnight at 110 °C. Cu–Ni–K/diatomite nano-catalysts were prepared by wetness impregnation method. Firstly, $Cu(NO_3)_2 \cdot 3H_2O$, $Ni(NO_3)_2 \cdot 6H_2O$ and KNO_3 were dissolved in ammonia solution with stirring, then the diatomite was dispersed in metallic ammonia solution. The resulting mixture was stirred at room temperature for 24 h, ultra-sonicated for another 3 h, followed by rotary evaporation to remove the solvent. Thereafter, it was dried at 110 °C overnight. The fully dried solid was calcined at 550 °C for 5 h and further reduced by mixed gas of H_2 (10%)/N_2 at 550 °C for 6 h. 3A MS was pretreated by calcining at 500 °C for 6 h, and then cooled down to room temperature, placed in a vacuum-desiccator for further use.

3.2. Catalyst Characterization

The surface area of the samples was detected in liquid N_2 by Brunauer-Emmett-Teller (BET) approaches using a Micromeritics ASAP 2010 (Micromeritics, Norcross, GA, USA) instrument. Thermogravimetric analyses (TGA) of samples were performed on a PerkinElmer Pyris Diamond SII thermal analyzer (high-purity N_2, 20 °C/min, PerkinElmer, Waltham, MA, USA). The morphologies of the samples were examined using a scanning electron microscopy (SEM) (JSM-5600LV system of JEOL (JEOL, Tokyo, Japan) equipped with an energy dispersive X-ray spectrometer (EDX) (JEOL, Tokyo, Japan) to check the components of the catalysts. The phase structure of the samples were determined by X-ray diffraction (XRD) (Rigaku Corporation, Tokyo, Japan) on a D/Max-IIIA power diffractometer using Cu (Kα) (0.15406 nm) radiation source. X-ray photoelectron spectrum (XPS) of the catalysts was obtained by ESCALAB 250 (ThermoFisher Scientific, Waltham, MA, USA) analyzer using the monochromatized Al (Kα) radiation source. Temperature programmed reduction (TPR)/Temperature programmed desorption (CO_2/NH_3-TPD) experiments of the samples were detected by Quantachrom ChemBET 3000 apparatus (Quantachrom Instruments, Boynton Beach, FL, USA) equipped with a thermal conductivity detector (TCD) [23].

3.3. Catalyst Evaluation

The evaluation of the catalysts was performed in a continuous tubular fixed-bed micro-gaseous reactor with 5 g of the fresh 15%(2Cu–Ni)-2%K_2O/diatomite (marked as 15CN2K/diatomite) catalyst and set mass ratio of the selected 3A MS as dehydrating agent. (3A/catalyst = 0/5, 5/5, 10/5, 15/5, 20/5, 25/5). The filling of the catalyst and 3A MS was stacked layer by layer alternatively and the top layer was the dehydrating agent, that is, each layer of the catalyst was sandwiched by two layers of dehydrating agent to ensure highly dehydrated. The methanol was bubbled into the reactor by N_2 and the molar ratio of CH_3OH and CO_2 was controlled by N_2 flux and the bubbling temperature (Scheme 1). The reaction was carried out at different temperatures, pressures and space velocity. The product was analyzed by GCMS-QP2010 Plus (SHIMADZU CORPORATION, Tokyo, Japan) and on-line GC (GC7890F) (TECHCOMP CORPORATE, Shang Hai, China) equipped with a flame ionization detector and thermal conductivity conductor. The final results were calculated by the following equations:

$$CH_3OH \text{ conversion } (\%) = \frac{[CH_3OH \text{ reacted}]}{[CH_3OH \text{ total}]} \times 100\% \tag{2}$$

$$DMC \text{ selectivity}(\%) = \frac{[DMC]}{[DMC + Byproduct]} \times 100\% \tag{3}$$

$$DMC \text{ yield}(\%) = CH_3OH \text{ conversion} \times DMC \text{ selectivity} \times 100\% \tag{4}$$

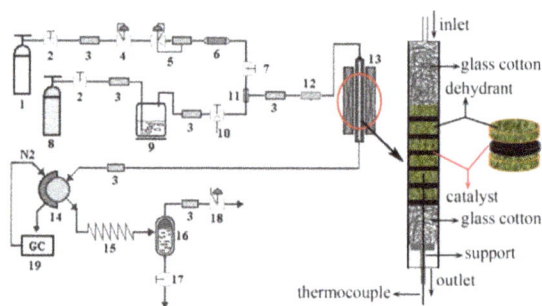

Scheme 1. Schematic diagram of the bubbling apparatus for direct synthesis of dimethyl carbonate (DMC).

4. Conclusions

An in-situ dehydrating process for the direct synthesis of DMC from methanol and CO_2 in a continuous flow fixed-bed bubbling micro-gaseous reactor was introduced, in which a sandwich structure of catalyst and 3A MS layer by layer was filled up. The effect of the mass ratio of 3A MS to catalyst, the dehydrating temperature and pressure, as well as the space velocity on the performance of the catalyst was investigated. The experimental results demonstrate that 3A MS can effectively dehydrate the catalytic reaction system at the optimal conditions of 120 °C, 1.2 MPa, with a GHSV of 600 h^{-1}. The in-situ dehydrating methodology enhances the methanol conversion and selectivity when compared with the dehydrating reaction system. Compared with the chemical dehydrating agents such as 2,2-dimethoxypropane [22], butylene oxide [24], etc., 3A MS is easily recyclable and they do not produce byproducts. Compared with the catalytic reaction system using a tin catalyst, a batch reactor separated from the recyclable dehydrating tube packed with 3A MS [25], this dehydrating reaction system using highly active K_2O-promoted Cu–Ni catalyst sandwiched by 3A MS was more preferable from the practical viewpoint. This report opens up a new way to circumventing the thermodynamic limitations of direct DMC synthesis, and would greatly prompt the researchers to design new dehydrating system for improving the efficiency of DMC synthesis directly from methanol and CO_2.

Author Contributions: D.H., Y.C., Y.M., S.W. and M.X. conceived and designed the experiments; D.H. and Y.C. performed the experiments and analyzed the data; Y.L. and S.W. contributed analysis tools. D.H. wrote this paper.

Acknowledgments: This research was funded by the National Natural Science Foundation of China (Grant No. 21376276, 21643002), Guangdong Province Sci & Tech Bureau (Grant No. 2017B090901003, 2016B010114004, 2016A050503001), Natural Science Foundation of Guangdong Province (Grant No. 2016A030313354), Guangzhou

Sci & Tech Bureau (Grant No. 201607010042) and Fundamental Research Funds for the Central Universities for financial support of this work. The authors would like to thank the above funding.

Conflicts of Interest: The authors declare no conflict of interest.

References

1. Fu, Z.W.; Meng, Y.Z. Research Progress in the Phosgene-Free and Direct Synthesis of Dimethyl Carbonate from CO_2 and Methanol. *Chem. Beyond Chlorine* **2016**, 363–385. [CrossRef]
2. Ono, Y. Catalysis in the production and reactions of dimethyl carbonate, an environmentally benign building block. *Appl. Catal. A* **1997**, *155*, 133–166. [CrossRef]
3. Zhou, Y.J.; Fu, Z.W.; Wang, S.J.; Xiao, M.; Han, D.M.; Meng, Y.Z. Electrochemical synthesis of dimethyl carbonate from CO_2 and methanol over carbonaceous material supported DBU in a capacitor-like cell reactor. *RSC Adv.* **2016**, *6*, 40010–40016. [CrossRef]
4. Han, M.S.; Lee, B.G.; Suh, I.; Kim, H.S.; Ahn, B.S.; Hong, S.I. Synthesis of dimethyl carbonate by vapor phase oxidative carbonylation of methanol over Cu-based catalysts. *J. Mol. Catal. A Chem.* **2001**, *170*, 225–234. [CrossRef]
5. Watanabe, Y.; Tatsumi, T. Hydrotalcite-type materials as catalysts for the synthesis of dimethyl carbonate from ethylene carbonate and methanol. *Microporous Mesoporous Mater.* **1998**, *22*, 399–407. [CrossRef]
6. Jessop, P.G.; Ikariya, T.; Noyori, R. Homogeneous catalysis in supercritical fluids. *Chem. Rev.* **1999**, *99*, 475–493. [CrossRef] [PubMed]
7. Puga, J.; Jones, M.E.; Molzahn, D.C.; Hartwell, G.E. Production of Dialkyl Carbonates from Alkanol, Carbon Monoxide and Oxygen Using a Novel Copper Containing Catalyst, or a Known Catalyst with a Chloro-Carbon Promoter. U.S. Patent 5,387,708, 7 February 1995.
8. Pimprom, S.; Sriboonkham, K.; Dittanet, P.; Fottinger, K.; Rupprechter, G.; Kongkachuichay, P. Synthesis of copper-nickel/SBA-15 from rice husk ash catalyst for dimethyl carbonate production from methanol and carbon dioxide. *J. Ind. Eng. Chem.* **2015**, *31*, 156–166. [CrossRef]
9. Fu, Z.W.; Zhong, Y.Y.; Yu, Y.H.; Long, L.Z.; Xiao, M.; Han, D.M.; Wang, S.J.; Meng, Y.Z. TiO_2-Doped CeO_2 Nanorod Catalyst for Direct Conversion of CO_2 and CH_3OH to Dimethyl Carbonate: Catalytic Performance and Kinetic Study. *ACS Omega* **2018**, *3*, 198–207. [CrossRef]
10. Ionescu, R.O.; Peres-Lucchese, Y.; Camy, S.; Tassaing, T.; Blanco, J.F.; Anne-Archard, G.; Riboul, D.; Condoret, J.S. Activation of CH_3OH and CO_2 by metallophthalocyanine complexes: Potential route to dimethyl carbonate. *Rev. Roum. Chim.* **2013**, *58*, 759–763.
11. Zheng, H.; Hong, Y.; Xu, J.; Xue, B.; Li, Y.X. Transesterification of ethylene carbonate to dimethyl carbonate catalyzed by CeO_2 materials with various morphologies. *Catal. Commun.* **2018**, *106*, 6–10. [CrossRef]
12. Prymak, I.; Prymak, O.; Wang, J.H.; Kalevaru, V.N.; Martin, A.; Bentrup, U.; Wohlrab, S. Phosphate functionalization of $CeO_2–ZrO_2$ solid solutions for the catalytic formation of dimethyl carbonate from methanol and carbon dioxide. *ChemCatChem* **2018**, *10*, 391–394. [CrossRef]
13. Zhang, Z.F.; Liu, Z.W.; Lu, J.; Liu, Z.T. Synthesis of dimethyl carbonate from carbon dioxide and methanol over $Ce_xZr_{1−x}O_2$ and [EMIM]Br/$Ce_{0.5}Zr_{0.5}O_2$. *Ind. Eng. Chem. Res.* **2011**, *50*, 1981–1988. [CrossRef]
14. Aouissi, A.; Al-Othman, Z.A.; Al-Amro, A. Gas-phase synthesis of dimethyl carbonate from methanol and carbon dioxide over $Co_{1.5}PW_{12}O_{40}$ keggin-type heteropolyanion. *Int. J. Mol. Sci.* **2010**, *11*, 1343–1351. [CrossRef] [PubMed]
15. Wu, X.L.; Xiao, M.; Meng, Y.Z.; Lu, Y.X. Direct synthesis of dimethyl carbonate on H_3PO_4 modified V_2O_5. *J. Mol. Catal. A Chem.* **2005**, *238*, 158–162. [CrossRef]
16. Chen, H.L.; Xiao, M.; Wang, S.J.; Han, D.M.; Meng, Y.Z. Direct synthesis of dimethyl carbonate from CO_2 and CH_3OH using 4A molecular sieve supported Cu-Ni bimetal catalyst. *Chin. J. Chem. Eng.* **2012**, *20*, 906–913.
17. Bian, J.; Xiao, M.; Wang, S.J.; Lu, Y.X.; Meng, Y.Z. Direct synthesis of DMC from CH_3OH and CO_2 over V-doped Cu–Ni/AC catalysts. *Catal. Commun.* **2009**, *10*, 1142–1145. [CrossRef]
18. Wang, X.J.; Xiao, M.; Wang, S.J.; Lu, Y.X.; Meng, Y.Z. Direct synthesis of dimethyl carbonate from carbon dioxide and methanol using supported copper (Ni, V, O) catalyst with photo-assistance. *J. Mol. Catal. A Chem.* **2007**, *278*, 92–96. [CrossRef]
19. Wu, X.L.; Meng, Y.Z.; Xiao, M.; Lu, Y.X. Direct synthesis of dimethyl carbonate (DMC) using Cu–Ni/VSO as catalyst. *J. Mol. Catal. A Chem.* **2006**, *249*, 93–97. [CrossRef]

20. Zhao, T.X.; Hu, X.B.; Wu, D.S.; Li, R.; Yang, G.Q.; Wu, Y.T. Direct synthesis of dimethyl carbonate from carbon dioxide and methanol at room temperature using imidazolium hydrogen carbonate ionic liquid as a recyclable catalyst and dehydrant. *ChemSusChem* **2017**, *10*, 2046–2052. [CrossRef] [PubMed]

21. Sakakura, T.; Saito, Y.; Okano, M.; Choi, J.C.; Sako, T. Selective conversion of carbon dioxide to dimethyl carbonate by molecular catalysis. *J. Org. Chem.* **1998**, *63*, 7095–7096. [CrossRef] [PubMed]

22. Kohno, K.; Choi, J.C.; Ohshima, Y.; Yili, A.; Yasuda, H.; Sakakura, T. Reaction of dibutyltin oxide with methanol under CO_2 pressure relevant to catalytic dimethyl carbonate synthesis. *J. Organomet. Chem.* **2008**, *693*, 1389–1392. [CrossRef]

23. Honda, M.; Suzuki, A.; Noorjahan, B.; Fujimoto, K.; Suzuki, K.; Tomishige, K. Low pressure CO_2 to dimethyl carbonate by the reaction with methanol promoted by acetonitrile hydration. *Catal. Commun.* **2009**, *30*, 4596–4598. [CrossRef] [PubMed]

24. Eta, V.; Maki-Arvela, P.; Leino, A.R.; Kordas, K.; Salmi, T.; Murzin, D.Y.; Mikkola, J.P. Synthesis of dimethyl carbonate from methanol and carbon dioxide: circumventing thermodynamic limitations. *Ind. Eng. Chem. Res.* **2010**, *49*, 9609–9617. [CrossRef]

25. Choi, J.C.; He, L.N.; Yasuda, H.; Sakakura, T. Selective and high yield synthesis of dimethyl carbonate directly from carbon dioxide and methanol. *Green Chem.* **2002**, *4*, 230–234. [CrossRef]

26. Han, D.M.; Chen, Y.; Wang, S.J.; Xiao, M.; Lu, Y.X.; Meng, Y.Z. Effect of alkali-doping on the performance of diatomite supported Cu-Ni bimetal catalysts for direct synthesis of dimethyl carbonate. *Catalysts* **2018**, *8*, 302. [CrossRef]

catalysts

MDPI

Article

Evolution of Water Diffusion in a Sorption-Enhanced Methanation Catalyst

Renaud Delmelle [1,*], Jasmin Terreni [2], Arndt Remhof [3], Andre Heel [1], Joris Proost [4] and Andreas Borgschulte [2]

1   Institute of Materials and Process Engineering (IMPE), Zurich University of Applied Sciences (ZHAW), Technikumstrasse 9, CH-8401 Winterthur, Switzerland; heel@zhaw.ch
2   Laboratory for Advanced Analytical Technologies, Swiss Federal Laboratories for Materials Science and Technology (Empa), Überlandstrasse 129, CH-8600 Dübendorf, Switzerland; jasmin.terreni@empa.ch (J.T.); Andreas.Borgschulte@empa.ch (A.B.)
3   Materials for Energy Conversion, Swiss Federal Laboratories for Materials Science and Technology (Empa), Überlandstrasse 129, CH-8600 Dübendorf, Switzerland; arndt.remhof@empa.ch
4   Institute of Mechanics, Materials and Civil Engineering (iMMC), Université catholique de Louvain, Place Sainte-Barbe 2, B-1348 Louvain-la-Neuve, Belgium; joris.proost@uclouvain.be
*   Correspondence: renaud.delmelle@zhaw.ch or deld@zhaw.ch; Tel.: +41-58-934-47-72; Fax: +41-58-935-71-83

Received: 28 May 2018; Accepted: 18 August 2018; Published: 21 August 2018

Abstract: Sorption-enhanced methanation has consequent advantages compared to conventional methanation approaches; namely, the production of pure methane and enhanced kinetics thanks to the application of Le Châtelier's principle. In this paper, we address the question of the long-term stability of a sorption-enhanced methanation catalyst-support couple: Ni nanoparticles on zeolite 5A. Compared to most conventional methanation processes the operational conditions of sorption-enhanced methanation are relatively mild, which allow for stable catalyst activity on the long term. Indeed, we show here that neither coking nor thermal degradation come into play under such conditions. However, a degradation mechanism specific to the sorption catalysis was observed under cyclic methanation/drying periods. This severely affects water diffusion kinetics in the zeolite support, as shown here by a decrease of the water-diffusion coefficient during multiple cycling. Water diffusion is a central mechanism in the sorption-enhanced methanation process, since it is rate-limiting for both methanation and drying.

Keywords: CO_2 methanation; catalysis; water sorption; water diffusion

1. Introduction

The key issues of intermittency and dispersion of primary renewable electricity sources find an answer in power-to-gas (P2G) strategies, where the excess of renewable electricity is converted into synthetic gas fuels, using hydrogen produced by water electrolysis as a primary reactant [1]. Renewable methane is produced from renewable hydrogen and carbon dioxide with a high efficiency (Sabatier reaction). This process can be implemented on a large scale [2], namely because renewable methane plants are based on relatively simple chemical reactors using earth-abundant Ni-based catalysts located near areas of renewable electricity production, abundant CO_2 sources such as biogas production and access to the natural gas grid.

Established methanation processes [3,4] involve fixed-bed, fluidized-bed and three-phase reactors with classic metal-support catalyst systems. Although other elements are studied at a fundamental level (e.g., Ru [5], Co [6], Mo [7] and Fe [8]), nickel remains the most suitable active metal (and by far the most widely used in commercial applications) when considering activity, selectivity and price [3,9]. However, the reaction temperature is above 250 °C, resulting in a thermodynamically limited conversion yield of less than 96% [10], which is further reduced by finite kinetics. Although the latter may be improved by appropriate catalysts, the thermodynamic limit can only be overcome by modifying thermodynamic conditions. A straightforward possibility is to increase the reaction pressure, which comes with an additional energy cost. The thermodynamic equilibrium is also shifted by the active removal of the water product from the catalyst reaction centers by adsorbing it in the water affine catalyst support such as zeolites in order to improve the reaction yield and kinetics (i.e., making use of the well-known Le Châtelier's principle) [11,12]. We have shown that this process runs optimally with specific parameters, notably with gas hourly space velocities (GHSV) which are lower by orders of magnitude than processes used in classic catalyzed methanation (i.e., on the order of $100 \, h^{-1}$ [13]). The reason for this is that, under certain gas flow conditions, moisture evolves in the reactor as a stable water front. As long as this front does not reach the reactor outlet (sorption-enhanced mode), pure methane is produced. When the zeolite support is saturated with water, a drying step is required. Specific criteria also come into play in terms of reactor temperature: an optimum needs to be found between the zeolite support water sorption capacity and the metal catalytic activity. In the case of Ni nanoparticles on zeolite 5A, the optimum is 300 °C at atmospheric pressure [13]. The need for a fine optimization of sorption-enhanced processes was also recently shown by numerical simulations [14].

Similar to normal methanation conditions, nickel is facing durability issues when used as a sorption catalyst that simultaneously affect the process performance, cost and environmental impact. Depending on the process considered and on the methanation conditions—GHSV, temperature, pressure, stoichiometry, and reactant purity—different deactivation mechanisms can potentially occur [3,15]: poisoning [16], fouling (coking) [17], thermal degradation [18], mechanical degradation (attrition, crushing) [19] and corrosion (leaching) [16,20]. Despite their differences, these mechanisms all affect the concentration of active sites on the catalyst surface, in turn lowering the apparent rate constant for methanation. The development of effective solutions to these deactivation issues is a crucial research topic for future applications [9,15]. Another emerging field is the synergy between the catalyst and its support [9,21], which can affect the system performance in many ways. There was concern that the Ni sorption catalyst is particularly sensitive to coking due to the low water concentration at the catalytically active centers [10,16,19]. We show in this publication that the degradation of the catalytic activity relevant during the reaction phase of a sorption catalyst is negligible at optimized conditions. We attribute this to the encapsulation of the Ni-particle in the inner of the zeolite structure, preventing irreversible carbon growth, but allowing exchange of reactants and products to and from the active surface, respectively. However, we observe a diminution of the water diffusion, a process relevant only in sorption catalysts during the regeneration (drying) phase.

In this paper, a focus is made on the long-term performance of the catalyst-support couple under optimum conditions for sorption-enhanced methanation, the latter of which were determined elsewhere [13], using nickel nanoparticles as the active metal and zeolite 5A as the support. The choice of zeolite 5A is justified elsewhere [22,23]. In short, the most important parameters for the choice of a sorption enhanced support are (i) the pore size, which allows manipulating the reaction path towards the desired reaction intermediates and products—because of the differences in product kinetic diameters, pore sizes ≥ 5 Å favor CH_4 formation while pore sizes ≤ 3 Å favor CO formation—and (ii) the water sorption capacity at the catalyst operation temperature, which determines the extent of Le Châtelier's effect.

The performance of this system was studied by means of a thermogravimetric method; i.e., with specimens under the gas stream in a magnetic suspension balance. This approach allows for both equilibrium and kinetic analyses through real-time monitoring of the specimen mass change. This reflects the evolution of the Sabatier reaction, because water is one of its products and is entirely adsorbed on the zeolite support as long as the reaction is sorption-enhanced. We also used this experimental approach to measure the equilibrium CO_2 and H_2O uptake capacity at conditions relevant for methanation. Methanation and drying have been studied over long periods of time, both in steady-state and multiple-cycling conditions. The evolution of the performance was then linked to the catalyst surface chemistry, crystal structure and water desorption kinetics.

2. Results and Discussion

The catalyst-support couple after reduction is as follows: Ni particles with sizes in the 20–30 nm range are present on the zeolite surface as well as in the bulk. The cube-like zeolite crystallites are typically between 2 and 5 μm in size, exhibiting flat facets over which the Ni particles are homogeneously distributed (see inset of Figure 1a). The fracture surface shown in Figure 1a is representative of the pellet bulk, as well as the surface, on which no noticeable difference was observed. The XRD data of the as-prepared specimen shown in Figure 1b confirms the Ni and zeolite crystallite sizes, as well as crystal structures.

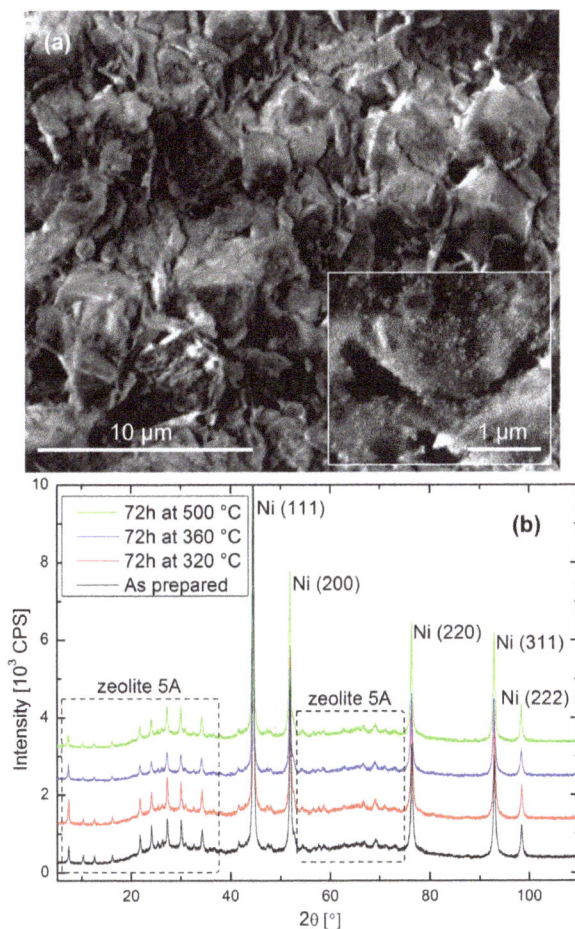

Figure 1. (**a**) Representative SEM image of the fracture surface of a Ni-impregnated 5A zeolite pellet after reduction. (**b**) X-ray powder diffraction patterns of the catalyst as prepared and after methanation under different conditions.

Continuous methanation experiments were carried out at different temperatures. In such experiments, the catalyst was simply subjected to a stoichiometric H_2/CO_2 ratio for a given time period. We will first focus on the catalyst performance, illustrated by the normalized CH_4/CO_2 ratio in Figure 2. One should note here that this ratio is only a relative estimation of the catalyst performance because part of the gas stream is bypassing the catalyst (see experimental section). It cannot be used to quantitatively assess the catalyst performance, e.g., by deducing the process methane yield. Although experiments up to 75 h were performed with in-situ gravimetric measurements, it should be noted that the operation time with in-situ IR measurements was currently limited to about 30 h. As explained above, the outlet gas line was heated to avoid water condensation, but still some water condensed in the optical cell. After about 30 h, the IR background was affected by the water signal and did not allow for a quantitative evaluation of the data. Alternative solutions are currently being studied in order to improve the setup.

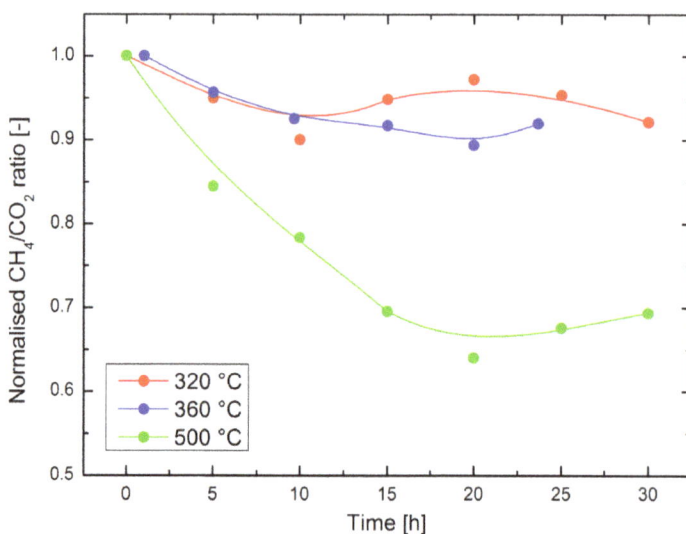

Figure 2. Evolution of the normalized CH_4/CO_2 ratio calculated from in-situ IR data during long-term methanation experiments at different temperatures. The solid lines are guides for the eye.

Relatively small performance drops were observed at 320 °C and at 360 °C, with a normalized CH_4/CO_2 ratio losing about 10% of its original value (i.e., at the start of the experiment, immediately after introduction of CO_2 into the system) in 30 h of operation in the optimum region for sorption-enhanced methanation. The performance drop was significantly higher at 500 °C, where a decrease of the CH_4/CO_2 ratio of 30% was observed after 30 h operation. This is an important result, which encourages the development of the still recent sorption-enhanced methanation strategy, because the latter exhibits mild operation conditions compared to most classic methanation catalysts [3,4]. One might indeed expect either Ni crystallite growth, catalyst poisoning or catalyst coking to be favored at higher operation temperatures. In the present study, poisoning can be ruled out because only pure commercial gases were used. Nickel crystallite growth is a possible scenario, especially at 500 °C. However, the XRD diffractograms in Figure 1 indicate there was no reduction in the peak width after 72 h of catalyst operation at any of the three temperatures considered here. This does not exclude fine changes in the microstructure of the nanoparticles. Studies on thermal degradation of classic methanation catalysts indicate that temperatures above 500 °C should generally be avoided [3]. The excellent stability of the catalyst activity in the optimal region for sorption-enhanced methanation (around 300 °C [13]) is encouraging. However, such materials are operated in alternating methanation/drying phases. The main challenge of the current study is the evaluation of the catalyst stability under such phases, which imply a focus on reversible water evolution in the zeolite support.

Figure 3a shows the evolution of the specimen mass as a function of time in a methanation/drying experiment performed at 360 °C. The steep early-stage kinetics is due to the sorption-enhanced effect, which results in fast methane production, and simultaneously, fast water uptake by the zeolite support. As water uptake approaches the zeolite saturation point, the sorption-enhanced effect becomes weaker, as shown by the slower late-stage kinetics. When the zeolite is saturated with water, the catalyst still produces methane, behaving like a classic methanation catalyst, where the system reflects the sole activity of the Ni nanoparticles. One must note here that the mass change shown in Figure 3a could not be exclusively the result of water adsorption by the zeolite support. As will be discussed later, the possibilities of specimen coking and CO_2 adsorption will be considered as well. Such phenomena would explain why the specimen mass was still significantly higher than the baseline acquired before methanation, even after 10 h drying.

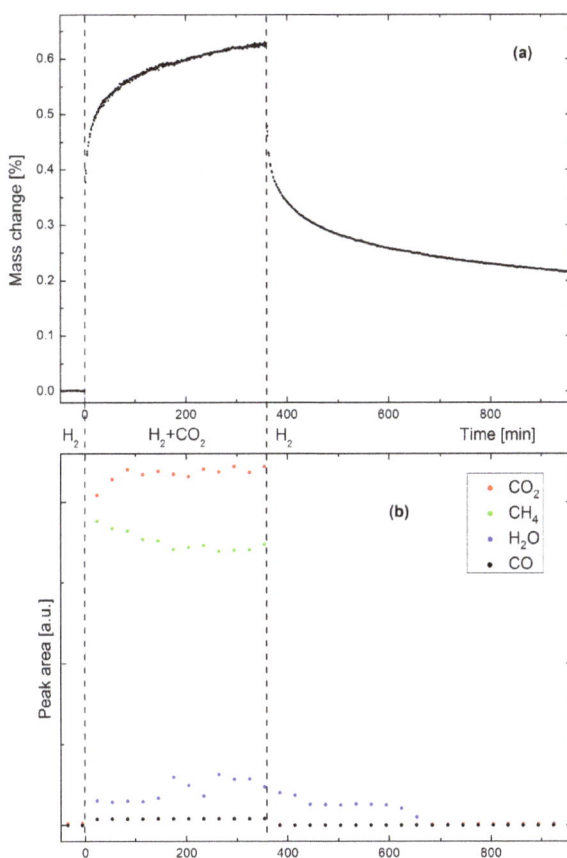

Figure 3. (a) Evolution of the mass of a Ni-impregnated 5A zeolite catalyst measured by a magnetic suspension balance during methanation (6 h) and drying (10 h) at 360 °C with stoichiometric H_2/CO_2 feed ratio. H_2 is always present in the inlet gas stream, and $t = 0$ s corresponds to CO_2 introduction into the system. **(b)** Evolution of the IR peak areas of CH_4, CO_2, H_2O, and CO in the outlet gas stream.

Figure 3b shows the peak areas of CO_2, CH_4, H_2O and CO as measured by IR spectrometry (one spectrum was recorded every 30 min). No IR peaks were observed before CO_2 introduction, as expected, since H_2 is not IR-active. As soon as the second reactant came into play, both CH_4 and H_2O products were observed, together with small amounts of CO, a well-known intermediate of the Sabatier reaction [22]. Small water peaks are readily observed because the system is not designed as a catalyst bed. It has previously been shown that pure methane can be produced only as long as the catalyst bed at the outlet consists of dry zeolite [13]. When the so-called water front reaches the outlet, a dynamic equilibrium state is reached between water on the zeolite surface and water in the gas stream. Here, water could escape the specimen at any time, even when the Sabatier reaction was still sorption-enhanced. Although no clear trends can be distinguished from the evolutions of the H_2O and CO peak areas, the trend observed in Figure 2 (i.e., a decreasing CH_4/CO_2 ratio) is already visible in a shorter experiment, such as in Figure 3b. When drying was initiated, water progressively desorbed from the zeolite, while the other gas species immediately dropped to zero. To corroborate the hypothesis of water being the origin of the mass change observed by gravimetry, we performed adsorption equilibrium measurements. Figure 4 shows the equilibrium uptake of CO_2 and water at fixed partial pressures and various temperatures, including the temperature range relevant for methanation. The CO_2 partial pressure of 200 mbar corresponded to that used in the methanation experiments, while the partial water pressure used for adsorption experiments was significantly lower than that expected to occur during methanation: at 50% CO_2 conversion, a water partial pressure of 200 mbar was reached, while the adsorption experiments were limited to a water partial pressure of 15.8 mbar for technical reasons (humidification of the carrier gas had to take place at room temperature). Still, the equilibrium water uptake exceeded that of CO_2 by a factor of five; the mass change observed during drying can thus be attributed to water. Apart from confirming the hypothesis of water being the origin of the mass change, the experiments yielded important information for the development of materials for sorption-enhanced methanation. The uptake curves match literature data [24] of the pure 5A zeolite over the full investigated temperature range, if scaled by a factor of around five (see Figure 4). This means that when anticipating performance data of sorption catalysts, it is possible to refer to reference data for the zeolite host with a "correction factor". This factor is due to the increased weight from the additional catalyst added to the sorption material (about 6 wt %), and partial blocking of adsorption sites due to the impregnation process (in a previous study on the same materials, we showed the effect of Ni loading on the catalyst BET surface area [22], e.g., the addition of 6 wt % Ni decreases the latter from 460 m^2g^{-1} to 330 m^2g^{-1}). The maximum uptake of the host sorption material determines the performance of the sorption catalyst. As pure zeolite 13X has a higher water uptake capacity than 5A catalysts, a better performance by it as the host material for the sorption catalyst is expected [13] and was found (the duration of the sorption enhanced mode is multiplied roughly by a factor of three when switching from 5A to 13X [23]). Furthermore, thermodynamic parameters, such as the heat of adsorption, relevant for the methanation and drying process remain unchanged. For energy balance calculations of the process, one may thus rely on reference data of the pure zeolites. As the energetics of the water-zeolite system remains unchanged, we can thus safely assume that the kinetic properties, such as the diffusion of the sorption catalyst, also reflect that of the pure host material.

Crank developed a model based on Fick's second law of diffusion for the evolution of water in porous solids [25], describing it as a water vapor diffusion process. The following expression accounts for the evolution of the average water content in a porous solid under the hypothesis of a uniform surface water concentration [26]:

$$\frac{m_t,}{m_{eq}} = 1 - \frac{6}{\pi^2} \sum_{n=1}^{\infty} \frac{1}{n^2} e^{-\frac{Dn^2\pi^2 t}{R^2}} \tag{1}$$

where m_t and m_{eq}, are the masses adsorbed at time t and at equilibrium, R is the average zeolite crystallite radius and D is the diffusion coefficient. Equation (1) is valid for both water adsorption and desorption [26] and can therefore be used to fit a water uptake mechanism such as sorption-enhanced methanation (Figure 3a) but also for zeolite drying [27], i.e., the two fundamental steps of the sorption-enhanced methanation process [13]. This means that in both cases, m_{eq} is the water uptake capacity of the zeolite support at the reaction temperature. In the case of methanation, it is the mass of water taken up by the zeolite; in the case of drying, it is the mass of water desorbed from the zeolite. A simplified expression of Equation (1) accounts for late-stage kinetics (long term asymptote, $m_t/m_{eq} > 0.75$) [26]:

$$\ln\left(1 - \frac{m_t}{m_{eq}}\right) = \ln\left(\frac{6}{\pi^2}\right) - \frac{\pi^2 D t}{R^2} \tag{2}$$

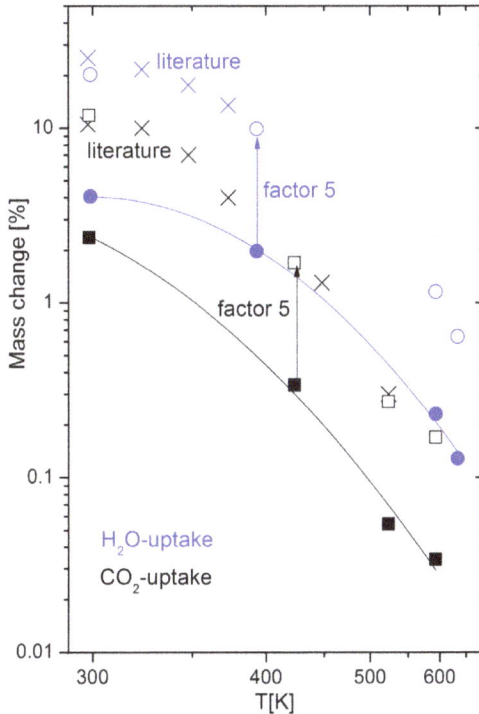

Figure 4. Equilibrium uptake of CO_2 and water at fixed partial pressures for various temperatures. The partial pressures are 200 mbar CO_2 and 15.8 mbar H_2O, respectively, in the carrier gas He. Literature data on pure zeolite 5A[24] is included for comparison. The literature data matches that for the sorption catalyst if scaled by a factor of five.

According to Equation (2), Figure 5 shows a logarithmic plot of the fractional mass uptake as a function of time from the data of Figure 3a. One can conclude from this plot that the methanation reaction is limited by water diffusion. However, on the one hand, choosing R as the zeolite crystallite radius does not result in realistic values of D. On the other hand, choosing R as the zeolite pellet size ($R = 0.075$ cm) results in $D = 7.6 \pm 0.1 \times 10^{-8}$ cm^2/s, in good agreement with literature [28]. This indicates that the rate-limiting step of sorption-enhanced methanation is water diffusion through the pellet bed rather than intracrystalline diffusion. The same analysis could be performed in the early-stage kinetics, for which a simplified expression of Equation (1) also exists [26], but this stage is too fast compared to the minimum time resolution of the magnetic suspension balance, so that no reliable quantitative analysis could be performed.

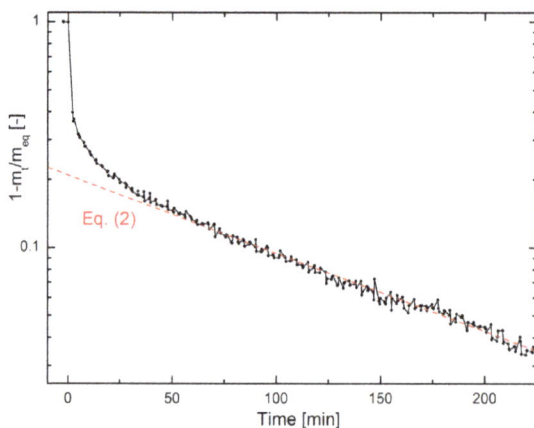

Figure 5. Evolution of the fractional mass uptake of a Ni-impregnated 5A zeolite catalyst during methanation at 360 °C with stoichiometric H_2/CO_2 feed ratio. The red line represents Equation (2).

The possibility of coking was studied by means of specimen melting in pure oxygen and subsequent quantitative infrared analysis (see experimental details). The methanation experiments described in Figure 1 were repeated with different operation times in order to evaluate the carbon concentration in the related specimens (shown in Figure 6). First, the carbon content increased roughly by a factor of four between the as-prepared specimens and the specimens that had been utilized for catalytic screening. Classical coking on the surface of the catalytic Ni particles is expected to increase with time as well as temperature [19]. The carbon content was influenced neither by the methanation time, nor by the reaction temperature, although the catalytic performance decreased at very high temperatures. It remained at an average of 0.11 ± 0.01 wt %; i.e., there was no significant carbon uptake by the sorption catalysts during catalysis. Therefore, the results from the performance tests, as well as the carbon analysis, demonstrate a very robust behavior of the catalysts, in particular at typical operation temperatures. This observation is in line with studies on coking of classic methanation catalysts [3], which indicate that coking is not an issue with operation temperatures under 500 °C.

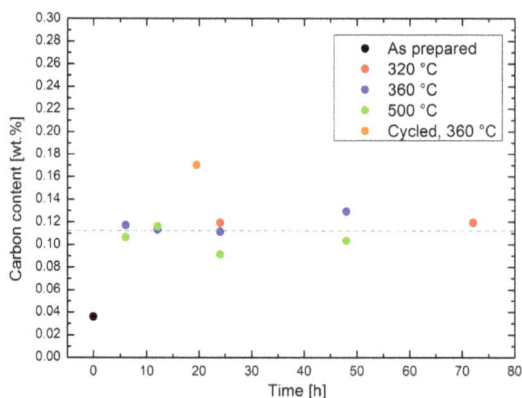

Figure 6. Gravimetric carbon content in Ni-impregnated 5A zeolite specimens subjected to continuous methanation experiments (red, blue and green dots, average of the latter shown by the grey line) and multiple methanation/drying cycles (orange).

Apart from water, zeolites adsorb CO_2 [29,30], CO, and various other carbon-containing side products, such as carbonates and higher hydrocarbons. For example, adsorbed CO_2 could remain in the zeolite pores even at zero CO_2 partial pressure [29], i.e., also during specimen transport in air between the methanation experiments and melting in the tubular oven, but this amount would not increase with reaction time. Since diffusion of CO_2 in the pores of zeolite 5A is relatively fast [30], other adsorbed carbon-containing intermediates may contribute to the total carbon content. Again, these compounds would not accumulate during the course of the reaction. With concentrations as low as 0.1 wt %, the chemical analysis of these compounds is a challenge. Diffusive reflective infrared Fourier transform spectroscopy (DRIFTS) can identify most of possible adsorbates on a catalyst surface. These DRIFTS measurements on Ni-zeolite showed the presence of three well-known reaction intermediates in the Sabatier process: CO as well as the formate and carbonate ions [22]. As the latter ones may also be formed on the zeolite surface, their diffusion to the catalytic Ni surfaces may be slightly slower than their formation, and thus cause accumulation in the zeolites. This hypothesis is underlined by the measurement of the carbon uptake of a cycled sorption catalyst. The sequence, consisting of a methanation step of 30 min and a drying step of 1 h, was repeated 39 times, for a total methanation time of 19.5 h. Here, the carbon content was higher than in the specimens that had undergone continuous methanation experiments, even with longer methanation times at the same temperature (see Figure 6). During drying, any carbonaceous species was no longer adsorbed in competition with water, which eased its adsorption in the zeolite, and may have blocked some pores. We believe that the special degradation phenomenon described above was taking place preferentially on a dry, rather than a wet, surface.

This phenomenon is quantitatively assessed here through a study of the zeolite drying kinetics of each methanation/drying cycle in the multiple cycling experiment described above. Figure 7a shows that the drying kinetics is clearly slowed from cycle to cycle. As in the methanation kinetics shown in Figure 5, the fractional mass loss can be fitted with Equation (2) in the late stage [27]. Here again, realistic values of the diffusion coefficient were obtained by using R as the pellet size, confirming the above hypothesis that water evolution in the impregnated zeolite is limited by diffusion in the pellets. Figure 7b shows the evolution of D as a function of the cycle number. The order of magnitude is in agreement with the above-calculated value of D in the methanation regime. Moreover, it decreased by about 40% after 39 cycles, indicating a deterioration of the water-diffusion properties. Such a trend was not visible in the continuous methanation experiments, which points again towards a degradation mechanism taking place preferentially during the drying process.

The degradation phenomenon investigated here is specific to sorption-enhanced methanation. It takes place during the drying phases, where the CO_2 adsorbed on the zeolite surface during the methanation phases is converted to intermediates, such as carbon monoxide, formates, and carbonates [22]. There is an indication that this adsorption takes place preferentially in dry conditions. This means that here, catalyst degradation is not associated to a decrease of the catalyst activity (i.e., catalyst deactivation). Despite the fact that the Ni nanoparticles maintain their activity throughout long sequences of methanation and drying, the diffusion properties of the water adsorbing zeolite host matrix are significantly affected, which in turns affects the process performance rather than the catalyst performance itself. Carbon containing adsorbates will affect the water adsorption kinetics and equilibrium [31] by a partial blocking of pores in the zeolite and hinder the Le Châtelier's effect in the sorption-enhanced methanation process. At higher temperatures, additional phenomena cannot be excluded: namely, the performance drop observed at 500 °C (see Figure 2) could be the result of hydriding of the Ni nanoparticles. Such particles embedded in a support could undergo important microstructural [32] and stress [33] effects, which in turn could strongly affect the distribution and mobility of hydrogen on the Ni surface. These possibilities will also be investigated in the future. An important outcome of the paper is the finding that the water diffusion path length is of the order of mm (Equation (2)). This is relevant for regeneration (drying) only: during methanation water diffuses from the catalytic reaction centers to the neighboring zeolite crystallites (i.e., of the order of

$10^{-8} \ldots 10^{-6}$ m); for regeneration, water has to leave the catalyst pellet, i.e., diffusion path lengths of 10^{-3} m (see Figure 8). This explains why the catalyst degradation affects the regeneration mode, while the catalytic performance during methanation is nearly unaffected.

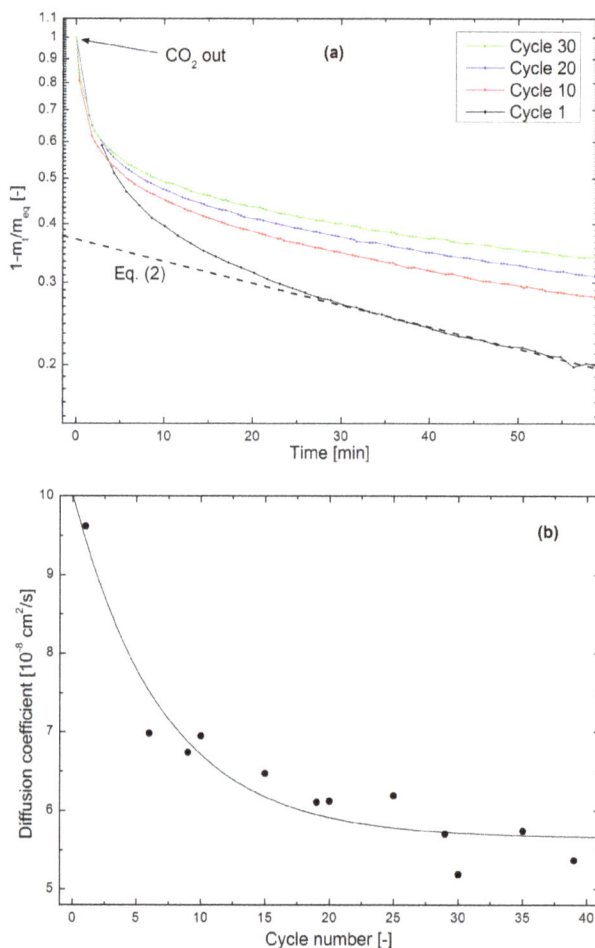

Figure 7. (a) Evolution of the fractional mass loss of a Ni-impregnated 5A zeolite catalyst during drying at 360 °C in a multiple cycling experiment. $t = 0$ s corresponds to CO_2 removal from the system. (b) Diffusion coefficient as a function of cycle number, calculated from fitting Equation (2) to the late-stage region of drying curves. The solid line is a guide to the eye.

Sorption-enhanced methanation enables the production of pure methane from CO_2 and hydrogen based on the active removal of water from the catalytically active (Ni-) centers. In a previous study, we showed that sorption-enhanced methanation catalysts indeed show better performances than commercial methanation catalysts with similar nanostructures operated in the same experimental conditions [11]. The reaction steps, which are enhanced, involve the formation of water, i.e., the detachment of oxygen from the intermediates (see, e.g., Shi et al. [34]). The subsequent reaction steps responsible for the hydrogenation of the carbons remain untouched. An unwanted side reaction is the formation of carbon deposits on the active catalyst, which will be enhanced if the hydrogenation of carbon is the rate-limiting step [19]. Thus, there are methanation concepts using a high water partial pressure preventing the accumulation of carbon on the catalyst [10]. The lower methanation yield of the single reaction is overcome by repeating the reaction in serial reactors [35]. Our reaction concept, based on a low water partial pressure, may thus be vulnerable to catalyst coking. We demonstrated in this paper that classical coking, i.e., carbon deposition on Ni, does not take place. However, due to a more complex materials system (sorption catalyst) and the specific operation conditions, an additional degradation mechanism was found to lead to reduced water diffusion in the sorbent. However, this problem can find solutions in the zeolite drying step, i.e., by making it a catalyst regeneration step as well. As a matter of fact, it was shown that drying in oxidizing atmosphere leads to better catalytic activity in the sorption-enhanced mode, even if it oxidizes the active metal at its surface (an increase of the duration of the sorption enhanced mode of 26% was observed when using air as a drying gas instead of hydrogen [23]). Oxidizing atmospheres can also be used in order to clean carbon compounds from the catalyst surface. Additionally, the simple fact that drying takes place in oxidizing conditions hinders the reduction of carbon oxides on the surface, and thereby the formation of coke. Of course, such oxidizing drying steps need to be balanced with reducing environments in order to limit oxidation of the active Ni surface.

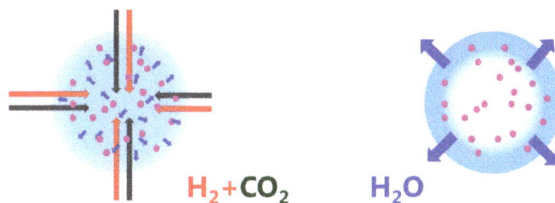

Figure 8. Sketch to illustrate the different length scale of diffusion: water diffusion is a local phenomenon during methanation (from the catalytic reaction centers to the neighboring zeolite crystallites, i.e., of the order of $10^{-8} \ldots 10^{-6}$ m); for regeneration, water has to leave the catalyst pellet, i.e., diffusion path lengths of 10^{-3} m.

3. Experimental Details

As stated above (see introduction), this study starts from a fully optimized process [13] with known catalyst microstructure and properties [11,23], from which the question of the long-term stability of the related catalyst arises. The metal-support system was prepared as follows: nickel nitrate hexahydrate ($Ni(NO_3)_2$, Sigma-Aldrich, St. Louis, MO, USA) was dissolved in deionized water. Pellets of zeolite 5A (Linde Type A, cylindrical shape, 1.5 mm diameter, 3 mm length) were immersed in this solution for 24 h at room temperature. Ni partly ion-exchanged with alkali/alkaline earth elements from the zeolite structure (zeolite 5A formula: $Ca_nNa_{12-2n}[(AlO_2)_{12}(SiO_2)_{12}] \cdot xH_2O$), while non-impregnated Ni could also remain in the zeolite pores after immersion. Therefore, the present synthesis is a combination of liquid ion exchange and impregnation. Washing of the pellets was omitted in order to avoid producing toxic Ni waste. The pellets were then dried at 100 °C for another 24 h, and reduced in hydrogen flow for 2 h at 650 °C. In these conditions, Ni atoms incorporated in the zeolite structure cluster and form nanoparticles on the surface as well as in the bulk, for a Ni loading of about 6 wt % [13].

This catalyst was characterized by the following methods: scanning electron microscopy (SEM) images of the zeolite surface were acquired using a Nova NanoSEM 230 FEI (Thermo Fischer Scientific, Waltham, MA, USA) with a 20 kV acceleration voltage. SEM observations were conducted at pellet surfaces and fracture surfaces without noticeable difference. The crystal structure of the specimens was investigated by X-ray powder diffraction (XRD) with a Bruker D8 diffractometer (Bruker, Billerica, MA, USA) with Cu-Kα radiation in a 2θ range of 5°–90° and a step size of 0.017°

(pattern PDF codes: 01-077-1335 (zeolite 5A) and 01-071-4740 4-850 (Ni)). Topas software (v. 5.0, Bruker AXS, Karlsruhe, Germany, 2014) was used for determination of crystallite sizes (Scherrer equation). Carbon concentrations were determined by melting the specimens in pure oxygen (with a slight overpressure compared to atmosphere) in a tubular oven at 1350 °C (IRF 1600 from SYLAB, Metz, France) and measuring the CO_2 emitted by carbon combustion in real time with nondispersive infrared spectrometry (Infrared Analyzer CSbox from SYLAB, with a resolution of 0.1 ppm).

The catalyst specimens were studied in-situ in a magnetic suspension balance (Rubotherm, Bochum, Germany) modified in order to encapsulate the specimen holder in a gas stream. This system is shown schematically in Figure 9. The outlet gas stream was connected to an infrared spectrometer (Alpha from Bruker, Billerica, MA, USA, equipped with an 8-cm gas cell, resolution: 0.9 cm^{-1}). The specimen could be heated up to 500 °C. The bucket-shaped specimen holder was micrometrer-sized meshed—any gas species could penetrate easily without undesirable flow effects. The gas connections were made with Swagelock tubing (Swagelock, Solon, OH, USA), and the gas flows controlled by thermal-based mass flow controllers (EL-FLOW Select series from Bronkhorst, Ruurlo, The Netherlands). The latter could be programmed with a LabVIEW interface. The outlet gas line was heated to 100 °C with a heating band in order to avoid water condensation in the optical cell. The main advantage of this system is the robustness of the magnetic suspension balance measurements. Reliable data could be acquired for days, either by continuous methane production or by repeated methanation/drying cycles. Apart from kinetic measurements, the system was used to acquire equilibrium data for the uptake of water and CO_2 as a function of temperature. For these experiments, the catalyst was exposed either to a 4:1 He/CO_2 gas mixture at 1 bar, or to humidified He gas corresponding to a water partial pressure of 15.8 mbar (50% humidity at 25 °C).

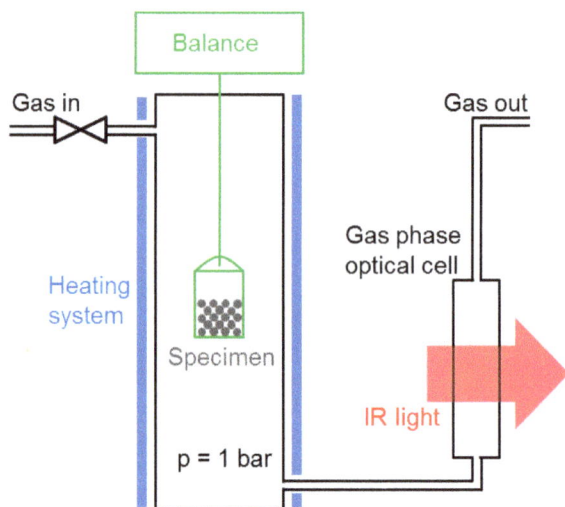

Figure 9. Schematic representation of the experimental setup used in this study for sorption-enhanced methanation and drying.

The specimen mass change during sorption-enhanced methanation and drying was monitored in this system, providing quantitative insight into the kinetics and equilibrium behaviors of the catalyst-support couple. The specimen mass was in the range 2.5–3 g. Both methanation and drying were performed at atmospheric pressure with a total flow of 250 mL/min and a stoichiometric H_2/CO_2 ratio. It should be noted that this setup is not a catalyst screening setup. No gas hourly space velocity, and thereby no catalyst turnover frequency, could be deduced from the total gas flow because the specimen holder was hanging in empty space, and therefore could not be considered as a catalyst bed. A significant CO_2 signal was always observed because the inlet gas species could easily bypass the specimen. Additionally, water could escape the specimen at any time and from any region. Consequently, the long-term performance of the catalyst will be discussed in terms of in-situ IR signal ratio between gas species, rather than absolute gas composition.

In the methanation experiments considered here, the specimens were permanently subjected to a H_2 flow; the CO_2 flow was then switched on to trigger methanation and simply switched off to trigger drying. In order to discuss the long-term catalyst performance in both classic and sorption-enhanced methanation, temperatures between 320 and 500 °C were considered; i.e., by screening the catalyst activity from the sorption-enhanced methanation optimum of 300 °C described by Borgschulte et al. [13] up to more commonly encountered temperatures for classic catalytic methanation [3,4].

4. Conclusions

Sorption-enhanced methanation catalysts allow for relatively mild operational conditions compared to conventional methanation; namely, in the case of Ni nanoparticles distributed on 5A zeolite, the active nanoparticles under operational conditions shows practically no long-term deactivation close to the sorption-enhanced methanation optimum, as coking and thermal degradation are excluded in the present study. However, sorption-enhanced methanation requires alternating methanation/drying periods. Water diffusion in the zeolite pellet bed is the rate-limiting step in both cases. A comparison of the equilibrium adsorption data of CO_2 and H_2O suggests that thermodynamic properties such as diffusion may be estimated from that of the pure zeolite, for which much more data exist. For equivalent methanation times, specimens screened under such cyclic treatments exhibited carbon contents about 55% higher than specimens subjected to continuous methanation conditions. This suggests a degradation phenomenon specific to sorption-enhanced methanation, which does not directly affect the catalytic activity of the active metal. In this sense, this degradation phenomenon is not a deactivation phenomenon. Reaction intermediates and products in the zeolite formed during methanation phases block pores in the sorbent during drying phases. The consequence of this was a decrease in the water diffusion coefficient of 40% observed after 39 cycles. The decrease in diffusion hardly affects the catalytic performance but has a considerable impact on the regeneration due to the much longer water diffusion path lengths. Further investigations are necessary in order to determine the mechanism of the reduction of diffusion, as well as the factors that influence the adsorption of reaction intermediates and products in the zeolite under dry conditions. Catalyst drying in oxidizing conditions is a good solution to this degradation phenomenon. More generally, we believe that the sorption-enhanced methanation strategy could be adapted to other power-to-fuel processes in the future (e.g., methanol synthesis, Fischer–Tropsch process).

Author Contributions: R.D. and A.B. conceived and designed the experiments, R.D. and J.T. performed the experiments, R.D. and A.B. analyzed the data, A.R. and J.P. performed materials analysis. R.D., A.H. and A.B. wrote and revised the paper.

Funding: This research was funded by the Swiss Federal Office of Energy (SFOE) and the 'Forschungs-, Entwicklungs- und Förderungsfonds der Schweizer Gaswirtschaft' (FOGA) through the SMARTCAT project (grant number SI/501130-01 and 0268) and by the Swiss National Science Foundation (SNSF) in the NRP70 programme 'Energy Turnaround'.

Conflicts of Interest: The authors declare no conflicts of interest.

References

1. Borgschulte, A. The Hydrogen Grand Challenge. *Front. Energy Res.* **2016**, *4*. [CrossRef]
2. Meylan, F.D.; Moreau, V.; Erkman, S. Material constraints related to storage of future European renewable electricity surpluses with CO_2 methanation. *Energy Policy* **2016**, *94*, 366–376. [CrossRef]

3. Rönsch, S.; Schneider, J.; Matthischke, S.; Schlüter, M.; Götz, M.; Lefebvre, J.; Prabhakaran, P.; Bajohr, S. Review on methanation—From fundamentals to current projects. *Fuel* **2016**, *166*, 276–296. [CrossRef]
4. Götz, M.; Lefebvre, J.; Mörs, F.; Koch, A.M.; Graf, F.; Bajohr, S.; Reimert, R.; Kolb, T. Renewable power-to-gas: A technological and economic review. *Renew. Energy* **2016**, *85*, 1371–1390. [CrossRef]
5. Kwak, J.H.; Kovarik, L.; Szanyi, J. CO_2 reduction on supported Ru/Al_2O_3 catalysts: Cluster size dependence of product selectivity. *ACS Catal.* **2013**, *3*, 2449–2455. [CrossRef]
6. Janlamool, J.; Praserthdam, P.; Jongsomjit, B. Ti-Si composite oxide-supported cobalt catalysts for CO_2 hydrogenation. *J. Nat. Gas Chem.* **2011**, *20*, 558–564. [CrossRef]
7. Aksoylu, A.E.; Misirli, Z.; Önsan, Z.I. Interaction between nickel and molybdenum in Ni-Mo/Al_2O_3 catalysts: I: CO_2 methanation and SEM-TEM studies. *Appl. Catal. A* **1998**, *168*, 385–397. [CrossRef]
8. Merkache, J.; Fechete, I.; Maamache, M.; Bernard, M.; Turek, P.; Al-Dalama, K.; Garin, F. 3D ordered mesoporous Fe-KIT-6 catalysts for methylcyclopentane (MCP) conversion and carbon dioxide (CO_2) hydrogenation for energy and environmental applications. *Appl. Catal. A* **2015**, *504*, 672–681. [CrossRef]
9. Aziz, M.A.A.; Jalil, A.A.; Triwahyono, S.; Ahmad, A. CO_2 methanation over heterogeneous catalysts: Recent progress and future prospects. *Green Chem.* **2015**, *17*, 2647–2663. [CrossRef]
10. Gao, J.; Wang, Y.; Ping, Y.; Hu, D.; Xu, G.; Gu, F.; Su, F. A thermodynamic analysis of methanation reactions of carbon oxides for the production of synthetic natural gas. *RSC Adv.* **2012**, *2*, 2358–2368. [CrossRef]
11. Borgschulte, A.; Gallandat, N.; Probst, B.; Suter, R.; Callini, E.; Ferri, D.; Arroyo, Y.; Erni, R.; Geerlings, H.; Züttel, A. Sorption enhanced CO_2 methanation. *Phys. Chem. Chem. Phys.* **2013**, *15*, 9620–9625. [CrossRef] [PubMed]
12. Walspurger, S.; Elzinga, G.D.; Dijkstra, J.W.; Sarić, M.; Haije, W.G. Sorption enhanced methanation for substitute natural gas production: Experimental results and thermodynamic considerations. *Chem. Eng. J.* **2014**, *242*, 379–386. [CrossRef]
13. Borgschulte, A.; Delmelle, R.; Duarte, R.B.; Heel, A.; Boillat, P.; Lehmann, E. Water distribution in a sorption enhanced methanation reactor by time resolved neutron imaging. *Phys. Chem. Chem. Phys.* **2016**, *18*, 17217–17223. [CrossRef] [PubMed]
14. Im, S.I.; Lee, K.B. Novel sorption-enhanced methanation with simultaneous CO_2 removal for the production of synthetic natural gas. *Ind. Eng. Chem. Res.* **2016**, *55*, 9244–9255. [CrossRef]
15. Moulijn, J.A.; van Diepen, A.E.; Katepijn, F. Catalyst deactivation: Is it predictable? What to do? *Appl. Catal. A* **2001**, *212*, 3–16. [CrossRef]
16. Argyle, M.D.; Bartholomew, C.H. Heterogeneous catalyst deactivation and regeneration: A review. *Catalysts* **2015**, *5*, 145–269. [CrossRef]
17. Bibby, D.M.; Howe, R.F.; McLellan, G.D. Coke formation in high-silica zeolites. *Appl. Catal. A* **1992**, *93*, 1–34. [CrossRef]
18. Zhang, G.; Sun, T.; Peng, J.; Wang, S.; Wang, S. A comparison of Ni/SiC and Ni/Al_2O_3 catalysed total methanation for production of synthetic natural gas. *Appl. Catal. A* **2013**, *462–463*, 75–81. [CrossRef]
19. Bartholomew, C.H. Mechanisms of catalyst deactivation. *Appl. Catal. A* **2001**, *212*, 17–60. [CrossRef]
20. Szabo, S.; Bakos, I. Corrosion accelerating surface catalysts. *Corros. Rev.* **2002**, *20*, 95–104. [CrossRef]
21. Yang, N.; Wang, R. Sustainable technologies for the reclamation of greenhouse gas CO_2. *J. Clean. Prod.* **2015**, *103*, 784–792. [CrossRef]
22. Borgschulte, A.; Callini, E.; Stadie, N.; Arroyo, Y.; Rossell, M.D.; Erni, R.; Geerlings, H.; Züttel, A.; Ferri, D. Manipulating the reaction path of the CO_2 hydrogenation reaction in molecular sieves. *Catal. Sci. Technol.* **2015**, *5*, 4613–4621. [CrossRef]
23. Delmelle, R.; Duarte, R.B.; Franken, T.; Burnat, D.; Holzer, L.; Borgschulte, A.; Heel, A. Development of improved nickel catalysts for sorption enhanced CO_2 methanation. *Int. J. Hydrog. Energy* **2016**, *41*, 20185–20191. [CrossRef]
24. Wang, Y.; LeVan, M.D. Adsorption equilibrium of carbon dioxide and water vapor on zeolites 5A and 13X and silica gel: Pure components. *J. Chem. Eng. Data* **2009**, *54*, 2839–2844. [CrossRef]
25. Crank, J. *The Mathematics of Diffusion*; Clarendon Press: Oxford, UK, 1970.
26. Ruthven, D.M. Diffusion in type A zeolites: New insights from old data. *Micropor. Mesopor. Mater.* **2012**, *162*, 69–79. [CrossRef]
27. Keey, R. *Drying of Loose and Particulate Materials*; Hemisphere Publishing Corporation: New York, NY, USA, 1992.

28. Turgman-Cohen, S.; Araque, J.C.; Hoek, E.M.V.; Escobedo, F.A. Molecular dynamics of equilibrium and pressure-driven transport properties of water through LTA-type zeolites. *Langmuir* **2013**, *29*, 12389–12399. [CrossRef] [PubMed]

29. Triebe, R.W.; Tezel, F.H. Adsorption of nitrogen, carbon monoxide, carbon dioxide and nitric oxide on molecular sieves. *Gas Sep. Purif.* **1995**, *9*, 223–230. [CrossRef]

30. Ruthven, D.M.; Lee, L.K.; Yucel, Y. Kinetics of non-isothermal sorption in molecular sieve crystals. *AIChE J.* **1980**, *26*, 16–23. [CrossRef]

31. Silva, J.A.C.; Mata, V.G.; Dias, M.M.; Lopes, J.C.B.; Rodrigues, A.E. Effect of coke in the equilibrium and kinetics of sorption on 5A molecular sieve zeolites. *Ind. Eng. Chem. Res.* **2000**, *39*, 1030–1034. [CrossRef]

32. Delmelle, R.; Amin-Ahmadi, B.; Sinnaeve, M.; Idrissi, H.; Pardoen, T.; Schryvers, D.; Proost, J. Effect of structural defects on the hydriding kinetics of nanocrystalline Pd thin films. *Int. J. Hydrog. Energy* **2015**, *40*, 7335–7347. [CrossRef]

33. Delmelle, R.; Michotte, S.; Sinnaeve, M.; Proost, J. Effect of internal stress on the hydriding kinetics of nanocrystalline Pd thin films. *Acta Mater.* **2013**, *61*, 2320–2329. [CrossRef]

34. Shi, C.; O'Grady, C.P.; Peterson, A.A.; Hansen, H.A.; Norskov, J. Modeling CO_2 reduction on Pt(111). *Phys. Chem. Chem. Phys.* **2013**, *15*, 7114–7122. [CrossRef] [PubMed]

35. Kopyscinski, J.; Schildhauer, T.J.; Biollaz, S.M.A. Production of synthetic natural gas (SNG) from coal and dry biomass—A technology review from 1950 to 2009. *Fuel* **2010**, *89*, 1763–1783. [CrossRef]

catalysts

Article

Effect of Alkali-Doping on the Performance of Diatomite Supported Cu-Ni Bimetal Catalysts for Direct Synthesis of Dimethyl Carbonate

Dongmei Han [1], Yong Chen [2], Shuanjin Wang [2], Min Xiao [2,*], Yixin Lu [3] and Yuezhong Meng [1,2,*]

[1] School of Chemical Engineering and Technology, Sun Yat-Sen University, Guangzhou 510275, China; handongm@mail.sysu.edu.cn

[2] The Key Laboratory of Low-Carbon Chemistry & Energy Conservation of Guangdong Province/State Key Laboratory of Optoelectronic Materials and Technologies, Sun Yat-Sen University, Guangzhou 510275, China; chenyong@lesso.com (Y.C.); wangshj@mail.sysu.edu.cn (S.W.)

[3] Department of Chemistry & Medicinal Chemistry Program, Office of Life Sciences, National University of Singapore, Singapore 117543, Singapore; chmlyx@nus.edu.sg

* Correspondence: stsxm@mail.sysu.edu.cn (M.X.); mengyzh@mail.sysu.edu.cn (Y.M.); Tel./Fax: +86-20-8411-4113 (Y.M.)

Received: 28 June 2018; Accepted: 24 July 2018; Published: 27 July 2018

Abstract: Alkali-adopted Cu-Ni/diatomite catalysts were designed and used for the direct synthesis of dimethyl carbonate (DMC) from carbon dioxide and methanol. Alkali additives were introduced into Cu-Ni/diatomite catalyst as a promoter because of its lower work function (Ni > Cu > Li > Na > K > Cs) and stronger electron-donating ability. A series of alkali-promoted Cu-Ni/diatomite catalysts were prepared by wetness impregnation method with different kind and different loading of alkali. The synthesized catalysts were fully characterized by means of X-ray diffraction (XRD), scanning electron microscope (SEM), temperature-programmed reduction (TPR), and NH_3/CO_2-TPD. The experimental results demonstrated that alkali adoption can significantly promote the catalytic activity of Cu–Ni bimetallic catalysts. Under the catalytic reaction conditions of 120 °C and 1.0 MPa; the highest CH_3OH conversion of 9.22% with DMC selectivity of 85.9% has been achieved when using 15%(2Cu-Ni) 2%Cs_2O/diatomite catalyst (CuO + NiO = 15 wt. %, atomic ratio of Cu/Ni = 2/1, Cs_2O = 2 wt. %).

Keywords: diatomite; alkali oxide; dimethyl carbonate; catalysis; carbon dioxide

1. Introduction

Carbon dioxide, the main greenhouse gas, can be converted into useful hydrocarbons rather than viewing it as waste emission [1]. Dimethyl carbonate (DMC), an environment-friendly building block, has attracted much attention as methylating and carbonylating agents, fuel additives, as well as polar solvents [2–5]. Direct catalytic synthesis of DMC from carbon dioxide and methanol has attracted much interest recently, which is industrially and environment-friendly compared to conventional commercial processes such as methanolysis of phosgene [6], ester exchange process [7,8], and gas-phase oxidative carbonylation of methanol [9]. Therefore, direct synthesis of DMC from CH_3OH and CO_2 is highly desired as it is environment-benign by nature [2]. However, highly efficient utilization of CO_2 is still a significant challenge because of its in-built thermodynamic stability and kinetic inert.

Many kinds of catalysts for the direct synthesis of DMC from CO_2, and CH_3OH has been reported, including organometallic compounds [10], potassium methoxide [11], ZrO_2, $Ce_{0.5}Zr_{0.5}O_2$, $H_3PW_{12}O_4$-$Ce_xTi_{1-x}O_2$, H_3PO_4-V_2O_5, $Co_{1.5}PW_{12}O_{40}$, and Rh/Al_2O_3 catalysts etc. [12–23]. Nevertheless, the performance of these catalysts has a long way to go. Therefore, a high-efficiency catalyst combined activation of CO_2 and methanol is under study. Some interesting investigations disclosed introducing a copper and nickel composite for the direct synthesis of DMC. S. H. Zhong et al. investigated the catalysts Cu-Ni/ZrO$_2$-SiO$_2$, Cu-Ni/MoO$_3$-SiO$_2$, and Cu-Ni/V$_2$O$_5$-SiO$_2$ for this catalytic reaction [24–27]. In our previous studies, X.L Wu et al. further optimized the preparation conditions and catalytic process of the Cu-Ni/VSO catalyst. In order to enhance the yield of DMC [28], X.J Wang et al. reported the similar Cu-(Ni,V,O)/SiO$_2$ catalysts with UV irradiation and pushed DMC yield close to 5% [29]. Following the progress of peers, a significant enhancement in catalytic activity and stability was achieved by J. Bian et al. [20–35].

Although the catalysts above-mentioned offer different advantages over others, considerable shortcomings still exist, such as a complicated preparation process, expensive support materials, and bleak prospects for large-scale preparation; moreover, some environment-destructive agents such as H$_2$SO$_4$, HF, and K$_2$MnO$_4$ were introduced during the process. Therefore, the investigation on the catalysts containing Cu-Ni bimetal, low-cost support, and high catalytic performance are much more meaningful from a practical point of view. In our previous work, a series of diatomite-immobilized Cu–Ni bimetallic nanocatalysts were prepared for the direct synthesis of dimethyl carbonate. The Cu-Ni bimetallic components supported on conductive carbon materials were reported [36]. It is found that the bimetallic composite is effectively alloyed and well immobilized inside or outside the pore of diatomite. Under the optimal conditions of 1.2 MPa and 120 °C, the prepared catalyst with loading of 15% exhibited the highest methanol conversion of 6.50% with DMC selectivity of 91.2% as well as more than 10 h lifetime [36].

Alkali additives are known to improve many industrially catalytic reactions such as ammonia and Fisher–Tropsch synthesis [37], CO oxidation and hydrogenation [38], and water-gas shift reaction [39,40]. It can induce a strong promotional effect on the performance of the catalysts such as enhanced activity and selectivity, suppression of undesirable reactions and improved catalyst stability. In this contribution, alkali-doped Cu-Ni/diatomite and pure Cu-Ni/diatomite bimetallic catalysts were prepared and characterized. The promotional effect of alkali on the dispersion, reduction, and activity of the catalyst are investigated in detail.

2. Results and Discussion

2.1. Fourier-Transform Infrared Spectroscopy (FTIR) Analysis of Diatomite

The FTIR spectrum of diatomite was recorded on an Analect RFX-65A type FTIR spectrophotometer with KBr matrix in region 450–4000 cm^{-1}. And the results are shown in Figure 1. The adsorption signature of -OH antisymmetric stretching vibration was exhibited at 3445 cm^{-1}. The -OH group was affected by the hydrogen bond coming from the adsorbed water on the surface, including pore water, and water bonded to the surface hydroxyl group. A broadband absorption signature at 1091 cm^{-1} and a shoulder absorption signature at 1200 cm^{-1} are attributed to Si-O antisymmetric stretching vibration. The absorption band at 471 cm^{-1} is attributed to the antisymmetric bending vibration of O-Si-O in SiO$_4$ tetrahedron. These spectrum features are consistent with amorphous SiO$_2$, which reflects the vibration characteristics of the SiO$_4$ tetrahedron of amorphous samples [41]. The hydroxyl groups existing on the surface and in the voids of the diatomite are extremely important for the infiltration of the precursor in and impregnated solution, and the adsorption and dispersion of the precursor on/in the diatomite.

2.2. Decomposition and Reduction Study of the Catalyst Precursor

The as-prepared Cu-Ni-M ammonia complex precursors were firstly investigated by thermogravimetric analysis (TGA) as shown in Figure 2. All the precursors exhibited clearly two-step decomposition. However, it is obvious that the decomposition temperature of alkali doped precursors is lower than that of the undoped precursor, which may result from the effect of distribution and induction of alkalis. It indicates a little lag around 300 °C at the end of the second decomposition step, which could be ascribed to the more difficult decomposition of alkali nitrate. Moreover, the decomposition completes much earlier with the increase of basicity and the decrease of potassium content for potassium doped precursors.

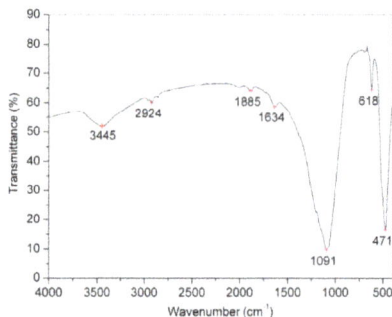

Figure 1. Fourier-transform infrared spectroscopy (FTIR) spectrum of diatomite.

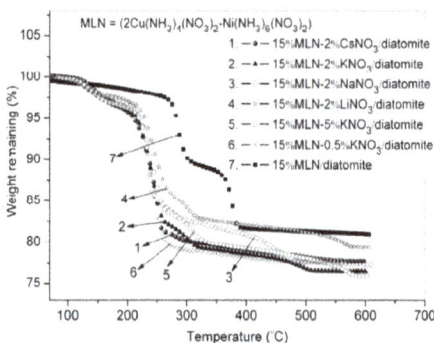

Figure 2. Thermogravimetric analysis (TGA) traces of catalyst precursor.

The calcined catalysts precursors were reduced by 5% H_2 purging (Figure 3). The CuO-NiO/diatomite shows two obvious overlapped reduction peaks of CuO (~303 °C) and NiO (~344 °C), respectively. The precursors doped with alkalis show one obvious combined reduction peak rather than overlapping combined reduction peaks, and all of them could be fully reduced below 400 °C. This may originate from the effect of distribution strongly promoted by alkali and the stronger reducibility induced by oxygen bridge bond between alkali and CuO/NiO (M-O-Cu/Ni) because of its much stronger adsorptive ability to H_2 [24]. Furthermore, it shifts to a slightly higher reduction temperature due to the less electron-accepted ability of the oxygen bridge bond with the increase of alkali basicity. The main reduction temperature increases from 309 °C to 325 °C with the increase of alkali basicity under the same content of alkali oxides (Figure 3b).

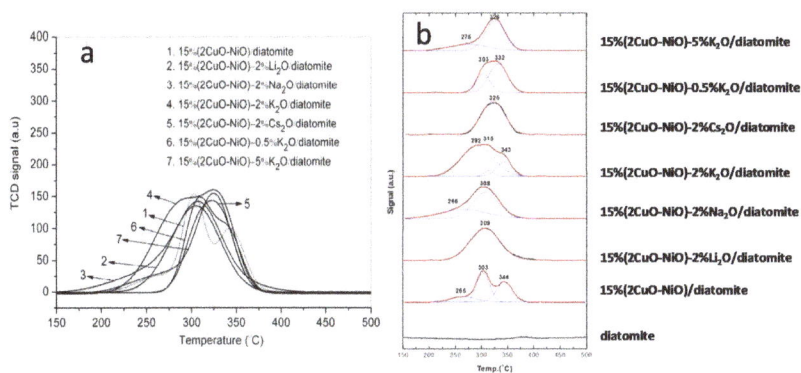

Figure 3. Temperature-programmed reduction (TPR) (**a**) and corresponding fitting curves (**b**) of calcined catalyst precursor.

2.3. Textural Investigation of the Catalyst

The powder X-Ray diffraction (XRD) study of alkali-doped and undoped Cu-Ni/diatomite are presented in Figure 4. All catalysts show four typical diffraction peaks of Cu-Ni alloy or Cu/Ni around 2θ value of 43.75 (111), 50.88 (200), 74.98 (220), and 91.13 (311) with very few diffraction peaks of CuO/NiO. Moreover, the weak diffraction peaks of 200, 220, and 311 become weaker and broader with increasing the basicity of alkali due to the alkali-promoted effect of lattice destruction and grain refinement, especially for cesium-doped catalyst, indicating that much stronger basicity of alkali is more favorable for the stabilization of the nano-particles. The same trend was also observed with increasing amount of K_2O, which resulted from the effect of potassium-promoted physical distribution. The effect of grain refinement may attribute to alkali as the nucleation agent for Cu-Ni precursor crystallization.

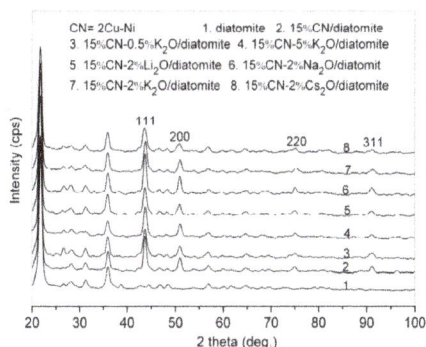

Figure 4. Powder X-ray diffraction (XRD) of the samples.

The morphology observation was conducted using scanning electron microscope (SEM) as shown in Figure 5. Figure 5a shows the natural diatomite and Figure 5b the treated diatomite, while Figure 5c shows the Cu-Ni/diatomite, Figure 5d–g shows the Cu-Ni/diatomite doped with 2% of LiO_2, Na_2O, K_2O, and Cs_2O in turn; the particle sizes of these catalysts decrease with the increase of the alkali basicity. Finally, the catalysts doped with 0.5% K_2O in Figure 5h and 5% K_2O in Figure 5i, compared with Figure 5f, evince that the particle size decreases with increasing the amount of K_2O, which is consistent with the result of XRD patterns. It indicates that the alkali doped Cu-Ni catalyst can facilitate the decomposition and reduction of Cu-Ni catalyst precursors at a much lower temperature. In addition, the alkali clusters dispersed in Cu-Ni crystallites could prevent the adjacent Cu-Ni grains from excessive growth at high temperature and thus stabilize the Cu-Ni crystallites. Consequently, more active sites of Cu-Ni surface containing alkali are exposed on the surface of reactant molecules, which favorites the catalytic reaction. A Transmission electron microscopy (TEM) image of 15%(2Cu-Ni)-2%K$_2$O/diatomite catalyst is shown in Figure 6; it can be seen that the catalyst particles are evenly dispersed on the diatomite support, and the particle size of the catalyst is about 20 nm. It could provide a high specific surface area and result in high utilization.

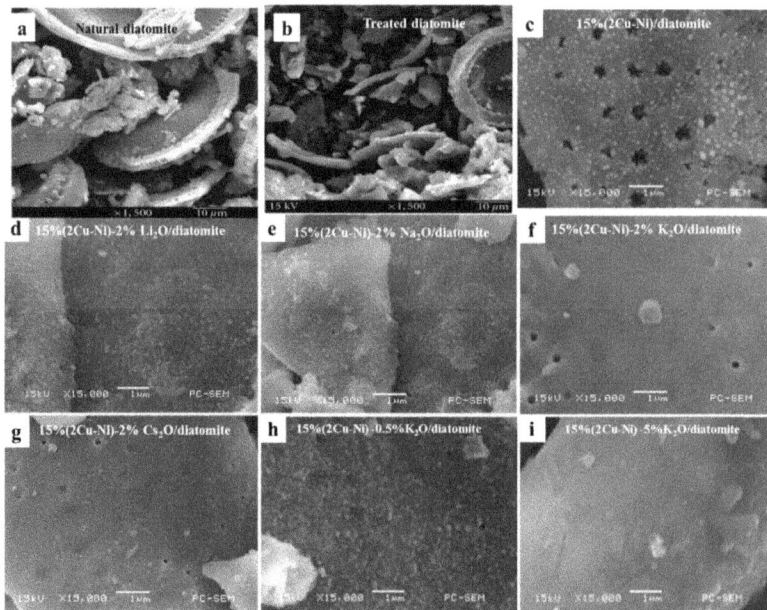

Figure 5. Scanning electron micrographic images of the samples.

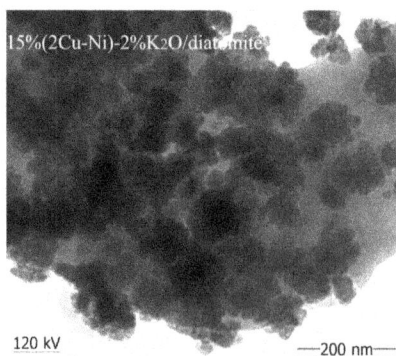

Figure 6. Transmission electron microscopy (EM) image of 15%(2Cu-Ni)-2%K$_2$O/diatomite catalyst.

2.4. Adsorptive Behavior of the Catalyst

The adsorptive properties of the catalyst samples were examined by CO_2-TPD (Temperature-Programmed Desorption) and NH_3-TPD, respectively. As shown in Figure 7, the catalysts doped with alkali exhibit much stronger CO_2 desorption than the undoped catalyst. In addition, the desorption peak slightly shifts to a higher temperature with the increase of alkali basicity, from a 157 °C increase to 223 °C, which increases from 199 °C to 231 °C with the increase of K_2O content, respectively. According to Figure 7 and Table 1, it gradually shows a trend of two peaks with the increase of alkali basicity. This indicates a greater ability of CO_2 activation, due to the well-dispersed alkali clusters in Cu-Ni. Figure 8 presents the NH_3-TPD curves of as-prepared catalyst samples. The samples doped with alkali exhibit a little higher desorption temperature than the undoped catalyst (from 186 °C increase to 207 °C), but which decrease slightly with the Cs_2O doping. The details could be seen from Figure 8 and Table 2. It demonstrates that the introduction of alkali into Cu-Ni composites intensified the NH_3 desorption. Presumably, this is due to the effect of alkali-promoted dispersion and alkali-induced electron distributions of Cu-Ni bimetal. This provides more unsaturated complex centers for the adsorption of NH_3 and the activation of CO_2.

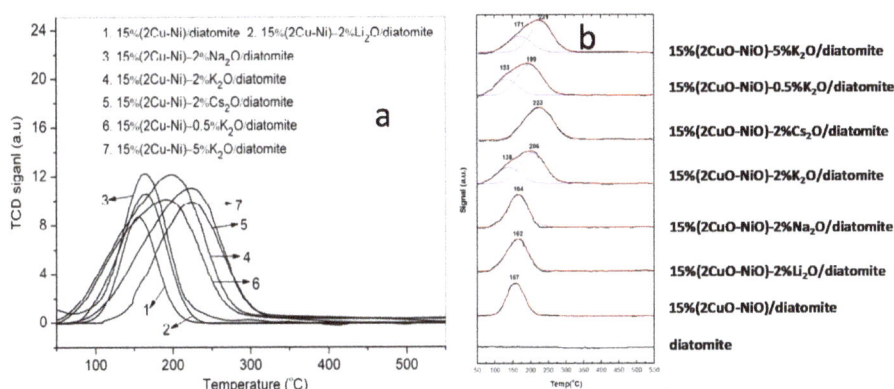

Figure 7. CO_2 Temperature-programmed desorption curves (**a**) and their fitting curves (**b**) of the samples.

Table 1. Quantification of the CO_2-TPD profiles of as-prepared catalysts.

Samples	T^a (°C)	Amount [b] (μmol gcat^{-1})	Total [c]
Diatomite	—	—	—
15%(2CuO-NiO)/diatomite	157	5.45	5.45
15%(2CuO-NiO)-2%Li_2O/diatomite	162	7.61	7.61
15%(2CuO-NiO)-2%Na_2O/diatomite	164	8.39	8.39
15%(2CuO-NiO)-2%K_2O/diatomite	138 206	4.75 9.79	14.54
15%(2CuO-NiO)-2%Cs_2O/diatomite	223	9.71	9.71
15%(2CuO-NiO)-0.5%K_2O/diatomite	133 199	3.75 8.09	11.84
15%(2CuO-NiO)-5%K_2O/diatomite	171 231	4.40 7.85	12.25

[a] Peak temperature of fitting curves; [b] Amount of absorption NH_3 (μmol) per gram catalyst according to each peak; [c] Total amount of absorption NH_3 (μmol) per gram catalyst.

Table 2. Quantification of the NH₃-TPD profiles of as-prepared catalysts.

Samples	T [a] (°C)	Amount [b] (μmol gcat $^{-1}$)	Total [c]
diatomite	—	—	—
15%(2CuO-NiO)/diatomite	134	1.87	5.64
	186	2.36	
	243	1.41	
15%(2CuO-NiO)-2%Li₂O/diatomite	200	6.60	6.60
15%(2CuO-NiO)-2%Na₂O/diatomite	202	9.96	9.96
15%(2CuO-NiO)-2%K₂O/diatomite	207	11.3	11.3
15%(2CuO-NiO)-2%Cs₂O/diatomite	185	11.2	11.2

[a] Peak temperature of fitting curves; [b] Amount of absorption NH₃ (μmol) per gram catalyst according to each peak;
[c] Total amount of absorption NH₃ (μmol) per gram catalyst.

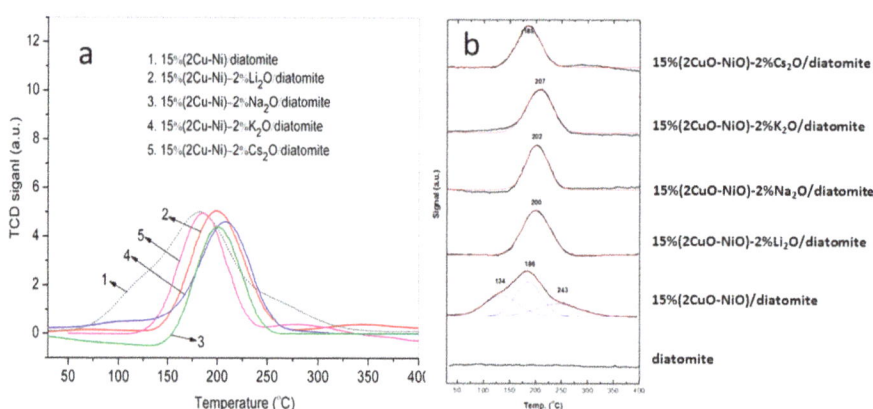

Figure 8. NH₃ Temperature-programmed desorption curves (**a**) and their fitting curves (**b**) of the samples.

2.5. Effect of Alkali on the Activity of Catalyst

As listed in Table 3, 15%(2Cu-Ni)/diatomite doped with different kinds and different amounts of alkali were prepared and studied. For the catalyst doped with Li₂O, the methanol conversion decreases from 6.11 to 2.77% with an increase of the Li₂O content. This is probably due to the destruction of the Cu-Ni alloy composite and the formation of the Cu-Li alloy. As for the catalysts doped with Na₂O, K₂O, and Cs₂O in turns, the catalytic activity increases with alkali dopant loading and reaches the highest value at 2wt % doping content (7.92 mol %). The highest doping content of 5wt % results in the lowest activity, likely owing to the excessive dopant surfacing on Cu-Ni, which can poison the active center of the Cu-Ni composites. Moreover, it seems that the catalytic activity is increased by increasing the basicity of dopant. It is believed that Na₂O, K₂O, and Cs₂O are solidified together with CuO-NiO during calcination when preparing catalysts. As a result, Na, K, and Cs can then immigrate into the lattice of the Cu-Ni alloy composite. Thus Cu-Ni-alkali (Na, K, and Cs) can be partly alloyed on the interface of the alkali oxide and Cu-Ni during the process of reduction. In conclusion, the introduction of alkalis into Cu-Ni lattice can promote the polarization of Cu-Ni lattice and speed up the electron transformation from Cu-Ni to CO₂, which in turn activates the reaction between methanol and CO₂. Compared with V-doped Cu-Ni catalyst [31], this method provides an effective and economic way for the direct synthesis of DMC, and would trigger much more interest in peer work.

Table 3. Influence of alkali loading on catalytic performance of 15% (2Cu-Ni)/diatomite.

Catalyst [a]	Methanol Conversion (mol %) [b,c]	DMC Selectivity (mol %) [c]	DMC Yield (mol %) [c]
15%(2Cu-Ni)/diatomite	6.50	91.2	5.93
15%(2Cu-Ni)-0.5%Li$_2$O/diatomite	6.11	88.2	5.39
15%(2Cu-Ni)-2%Li$_2$O/diatomite	5.68	83.2	4.73
15%(2Cu-Ni)-5%Li$_2$O/diatomite	2.77	85.1	2.36
15%(2Cu-Ni)-0.5%Na$_2$O/diatomite	6.68	83.3	5.56
15%(2Cu-Ni)-2%Na$_2$O/diatomite	7.02	84.5	5.93
15%(2Cu-Ni)-5%Na$_2$O/diatomite	3.97	81.7	3.24
15%(2Cu-Ni)-0.5%K$_2$O/diatomite	6.81	89.2	6.08
15%(2Cu-Ni)-2%K$_2$O/diatomite	7.55	90.3	6.82
15%(2Cu-Ni)-5%K$_2$O/diatomite	3.68	84.8	3.12
15%(2Cu-Ni)-0.5%Cs$_2$O/diatomite	7.17	90.7	6.50
15%(2Cu-Ni)-2%Cs$_2$O/diatomite	9.22	85.9	7.92
15%(2Cu-Ni)-5%Cs$_2$O/diatomite	5.65	80.4	4.54

[a] Molar ratio of CuO/NiO is 2/1, all metal contents are calculated by mass of corresponding metal oxide; [b] DMC yield is calculated based on the amount of methanol; [c] Reaction conditions: 120 °C; 1.0 Mpa; CO$_2$ flux (15 mL/min).

3. Experimental

3.1. Catalyst Preparation

Cu-Ni-M/diatomite (M = Li, Na, K, Cs) nanocatalysts were prepared by the wetness impregnation method. Firstly Cu(NO$_3$)$_2$·3H$_2$O, Ni(NO$_3$)$_2$·6H$_2$O and alkali nitrate were dissolved in ammonia solution with stirring, and then natural diatomite was dispersed in metallic ammonia solution. The resulting mixture was stirred at room temperature for 24 h, ultrasonicated for another 3 h, followed by rotavaporation to remove the solvent. Thereafter, it was dried at 110 °C overnight. The fully dried solid was calcining at 500 °C for 3 h and further reduced by 5% H$_2$/N$_2$ mixture at 550 °C for 6 h.

3.2. Catalyst Characterization

TGA of samples were performed on a PerkinElmer Pyris Diamond SII thermal analyzer (high-purity N$_2$, 20 °C/min). The morphologies of the samples were characterized using a SEM (JSM-5600LV, JEOL, Tokyo, Japan) equipped with an EDX to check the components of the catalysts. The phase structure of the samples was determined by XRD on a D/Max-IIIA power diffractometer (Rigaku Corporation, Tokyo, Japan) using Cu (Kα) (0.15406 nm) radiation source. Temperature programmed reduction (TPR) and Temperature programmed desorption of ammonia (NH$_3$-TPD)/carbon dioxide (CO$_2$-TPD) experiments of the samples were detected by Quantachrom ChemBET 3000 apparatus (Quantachrom Instruments, Boynton Beach, FL, USA) equipped with a thermal conductivity detector (TCD) [31].

The evaluation of the catalysts was performed in a continuous tubular fixed-bed micro-gaseous reactor with 2 g of the fresh catalyst and set molar ratio of CH$_3$OH bubbled into the reactor by CO$_2$ (30 mL/min flux). It was carried out under set conditions of 120 °C and 1.2 MPa. The products were analyzed by on-line GC (GC7890F) (TECHCOMP CORPORATE, Shang Hai, China) equipped with a flame ionization detector and GCMS-QP2010 Plus (SHIMADZU CORPORATION, Tokyo, Japan). The final results were calculated by the following Equations (1)–(3):

$$CH_3OH \; conversion(mol \, \%) = \frac{[CH_3OH]_{reacted}}{[CH_3OH]_{total}} \times 100\% \tag{1}$$

$$DMC \; selevtivity(mol \, \%) = \frac{[DMC]}{[DMC] + [Byproduct]} \times 100\% \tag{2}$$

$$DMC \; yield(mol \, \%) = CH_3OH \; conversion \times DMCselevtivity \tag{3}$$

4. Conclusions

Based on the SEM, TPR/TPD investigation of the activity and stability evaluation of the alkali-doped catalyst, we can conclude that the incorporation of alkali is conducive to the preparation of the catalysts precursor by decreasing the decomposition and reduction temperatures, which is favorable for the formation of a nano-scale dispersion of bimetalic particles on the surface of supports. The well-dispersed characteristic in turn endows the catalyst with more lattice drawbacks and a polarized Cu-Ni lattice. This effect becomes more obvious with increasing the basicity of alkali. The catalytic activity of the alkali-promoted catalyst is enhanced with the increase of alkali basicity, except lithium oxide, indicating alkali doping can significantly improve the catalytic efficiency of Cu-Ni composites. This preliminary study provides a new practical way to improve the efficiency of DMC synthesis, which will promote related research and peer distribution in this hot research area.

Author Contributions: D.H., Y.C., Y.M., S.W. and M.X. conceived and designed the experiments; D.H. and Y.C. performed the experiments and analyzed the data; Y.L. and S.W. contributed analysis tools. D.H. wrote this paper.

Acknowledgments: This research was funded by the National Natural Science Foundation of China (Grant No. 21376276, 21643002), Guangdong Province Sci & Tech Bureau (Grant No. 2017B090901003, 2016B010114004, 2016A050503001), Natural Science Foundation of Guangdong Province (Grant No. 2016A030313354), Guangzhou Sci & Tech Bureau (Grant No. 201607010042) and Fundamental Research Funds for the Central Universities for financial support of this work. The authors would like to thank the above funding.

Conflicts of Interest: The authors declare no conflict of interest.

References

1. Fu, Z.W.; Meng, Y.Z. research progress in the phosgene-free and direct synthesis of dimethyl carbonate from CO_2 and methanol. In *Chemistry beyond Chlorine*; Springer International Publishing: Cham, Switzerland, 2016; Chapter 13; pp. 363–386.
2. Zhou, Y.J.; Fu, Z.W.; Wang, S.J.; Xiao, M.; Han, D.M.; Meng, Y.Z. Electrochemical synthesis of dimethyl carbonate from CO_2 and methanol over carbonaceous material supported DBU in a capacitor-like cell reactor. *RSC Adv.* **2016**, *6*, 40010–40016. [CrossRef]
3. Ono, Y. Catalysis in the production and reactions of dimethyl carbonate, an environmentally benign building block. *Appl. Catal. A Gen.* **1997**, *155*, 133–166. [CrossRef]
4. Santos, B.A.V.; Silva, V.M.T.M.; Loureiro, J.M.; Rodrigues, A.E. Review for the direct synthesis of dimethyl carbonate. *ChemBioEng Rev.* **2015**, *1*, 214–229. [CrossRef]
5. Tundo, P.; Selva, M. The chemistry of dimethyl carbonate. *Accounts Chem. Res.* **2002**, *35*, 706–716. [CrossRef]
6. Jessop, P.G.; Ikariya, T.; Noyori, R. Homogeneous catalysis in supercritical fluids. *Chem. Rev.* **1999**, *99*, 475–493. [CrossRef] [PubMed]
7. Han, M.S.; Lee, B.G.; Suh, I.; Kim, H.S.; Ahn, B.S.; Hong, S.I. Synthesis of dimethyl carbonate by vapor phase oxidative carbonylation of methanol over Cu-based catalysts. *J. Mol. Catal. A Chem.* **2001**, *170*, 225–234. [CrossRef]
8. Watanabe, Y.; Tatsumi, T. Hydrotalcite-type materials as catalysts for the synthesis of dimethyl carbonate from ethylene carbonate and methanol. *Microporous Mesoporous Mater.* **1998**, *22*, 399–407. [CrossRef]
9. Puga, J.; Jones, M.E.; Molzahn, D.C.; Hartwell, G.E. *Production of Dialkyl Carbonates from Alkanol, Carbon Monoxide and Oxygen Using a Novel Copper Containing Catalyst, or a Known Catalyst with a Chloro-Carbon Promoter*; Dow Chemical Company: Midland, MI, USA, 1995.
10. Jia, G.; Gao, Y.F.; Zhang, W.; Wang, H.; Gao, Z.Z.; Li, C.H.; Liu, J.R. Metal-organic frameworks as heterogeneous catalysts for electrocatalytic oxidative carbonylation of methanol to dimethyl carbonate. *Electrochem. Commun.* **2013**, *34*, 211–214. [CrossRef]
11. Cai, Q.H.; Lu, B.; Guo, L.J.; Shan, Y.K. Studies on synthesis of dimethyl carbonate from methanol and carbon dioxide. *Catal. Commun.* **2009**, *10*, 605–609. [CrossRef]
12. Akune, T.; Morita, Y.; Shirakawa, S.; Katagiri, K.; Inumaru, K. ZrO_2 nanocrystals as catalyst for synthesis of dimethylcarbonate from methanol and carbon dioxide: Catalytic activity and elucidation of active sites. *Langmuir* **2018**, *34*, 23–29. [CrossRef] [PubMed]

13. Aouissi, A.; Al-Othman, Z.A.; Al-Amro, A. Gas-phase synthesis of dimethyl carbonate from methanol and carbon dioxide over $Co_{1.5}PW_{12}O_{40}$ keggin-type heteropolyanion. *Int. J. Mol. Sci.* **2010**, *11*, 1343–1351. [CrossRef] [PubMed]

14. Zhang, Z.F.; Liu, Z.T.; Liu, Z.W.; Lu, J. DMC formation over $Ce_{0.5}Zr_{0.5}O_2$ prepared by complex-decomposition method. *Catal. Lett.* **2009**, *129*, 428–436. [CrossRef]

15. La, K.W.; Jung, J.C.; Kim, H.; Baeck, S.H.; Song, I.K. Effect of acid-base properties of $H_3PW_{12}O_{40}/CexTi1-xO_2$ catalysts on the direct synthesis of dimethyl carbonate from methanol and carbon dioxide: A TPD study of $H_3PW_{12}O_{40}/CexTi1-xO_2$ catalysts. *J. Mol. Catal. A Chem.* **2007**, *269*, 41–45. [CrossRef]

16. Wu, X.L.; Xiao, M.; Meng, Y.Z.; Lu, Y.X. Direct synthesis of dimethyl carbonate on H_3PO_4 modified V_2O_5. *J. Mol. Catal. A Chem.* **2005**, *238*, 158–162. [CrossRef]

17. Almusaiteer, K. Synthesis of dimethyl carbonate (DMC) from methanol and CO_2 over Rh-supported catalysts. *Catal. Commun.* **2009**, *10*, 1127–1131. [CrossRef]

18. Bansode, A.; Urakawa, A. Continuous DMC synthesis from CO_2 and methanol over a CeO_2 catalyst in a fixed bed reactor in the presence of a dehydrating agent. *ACS Catal.* **2014**, *4*, 3877–3880. [CrossRef]

19. Stoian, D.; Medina, F.; Urakawa, A. Improving the stability of CeO_2 catalyst by rare earth metal promotion and molecular insights in the dimethyl carbonate synthesis from CO_2 and methanol with 2-cyanopyridine. *ACS Catal.* **2018**, *8*, 3181–3193. [CrossRef]

20. Pimprom, S.; Sriboonkham, K.; Dittanet, P.; Föttinger, K.; Rupprechter, G.; Kongkachuichay, P. Synthesis of copper–nickel/SBA-15 from rice husk ash catalyst fordimethyl carbonate production from methanol and carbon dioxide. *J. Ind. Eng. Chem.* **2015**, *31*, 156–166. [CrossRef]

21. Kang, K.H.; Lee, C.H.; Kim, D.B.; Jang, B.; Song, I.K. NiO/CeO_2–ZnO Nano-catalysts for direct synthesis of dimethyl carbonate from methanol and carbon dioxide. *J. Nanosci. Nanotechnol.* **2014**, *14*, 8693–8698. [CrossRef] [PubMed]

22. Tamboli, A.H.; Chaugule, A.A.; Kim, H. Catalytic developments in the direct dimethyl carbonate synthesis from carbon dioxide and methanol. *Chem. Eng. J.* **2017**, *323*, 530–544. [CrossRef]

23. Devaiah, D.; Reddy, L.H.; Park, S.E.; Reddy, B.M. Ceria–zirconia mixed oxides: Synthetic methods and applications. *Catal. Rev.* **2018**, *60*, 177–277. [CrossRef]

24. Li, H.S.; Zhong, S.H.; Wang, J.W.; Xiao, X.F. Effect of K_2O on adsorption and reaction of CO_2 and CH_3OH over Cu-Ni/ZrO_2-SiO_2 catalyst for synthesis of dimethyl carbonate. *Chin. J. Catal.* **2001**, *22*, 353–357.

25. Zhong, S.H.; Li, H.S.; Wang, J.W.; Xiao, X.F. Study on Cu-Ni/ZrO_2-SiO_4 catalyst for direct synthesis of dimethyl carbonate from CO_2 and CH_3OH. *J. Catal.* **2000**, *21*, 117–120.

26. Zhong, S.H.; Li, H.S.; Wang, J.W.; Xiao, X.F. Study on Cu-Ni/MoO_3-SiO_2 catalyst for the direct synthesis of dimethyl carbonate from carbon dioxide and methanol. *Pet. Process. Petrochem.* **2000**, *6*, 51–55.

27. Zhong, S.H.; Li, H.S.; Wang, J.W.; Xiao, X.F. Cu-Ni/V_2O_5-SiO_2 catalyst for the direct synthesis of dimethyl carbonate from carbon dioxide and methanol. *Acta Phys. Chim. Sin.* **2000**, *16*, 226–231.

28. Wu, X.L.; Meng, Y.Z.; Xiao, M.; Lu, Y.X. Direct synthesis of dimethyl carbonate (DMC) using Cu-Ni/VSO as catalyst. *J. Mol. Catal. A Chem.* **2006**, *249*, 93–97. [CrossRef]

29. Wang, X.J.; Xiao, M.; Wang, S.J.; Lu, Y.X.; Meng, Y.Z. Direct synthesis of dimethyl carbonate from carbon dioxide and methanol using supported copper (Ni, V, O) catalyst with photo-assistance. *J. Mol. Catal. A Chem.* **2007**, *278*, 92–96. [CrossRef]

30. Bian, J.; Xiao, M.; Wang, S.J.; Wang, X.J.; Lu, Y.X.; Meng, Y.Z. Highly effective synthesis of dimethyl carbonate from methanol and carbon dioxide using a novel copper-nickel/graphite bimetallic nanocomposite catalyst. *Chem. Eng. J.* **2009**, *147*, 287–296. [CrossRef]

31. Bian, J.; Xiao, M.; Wang, S.J.; Lu, Y.X.; Meng, Y.Z. Direct synthesis of DMC from CH_3OH and CO_2 over V-doped Cu-Ni/AC catalysts. *Catal. Commun.* **2009**, *10*, 1142–1145. [CrossRef]

32. Chen, H.L.; Wang, S.J.; Xiao, M.; Han, D.M.; Lu, Y.X.; Meng, Y.Z. Direct synthesis of dimethyl carbonate from CO and CHOH using 0.4 nm molecular sieve supported Cu-Ni bimetal catalyst. *Chin. J. Chem. Eng.* **2012**, *20*, 906–913. [CrossRef]

33. Zhou, Y.J.; Wang, S.J.; Xiao, M.; Han, D.M.; Lu, Y.X.; Meng, Y.Z. Formation of dimethyl carbonate on nature clay supported bimetallic Cu–Ni catalysts. *J. Clean. Prod.* **2014**, *103*, 925–933. [CrossRef]

34. Zhang, M.; Alferov, K.A.; Xiao, M.; Han, D.M.; Wang, S.J.; Meng, Y.Z. Continuous dimethyl carbonate synthesis from CO_2 and methanol using Cu-Ni@VSiO as catalyst synthesized by a novel sulfuration method. *Catalysts* **2018**, *8*, 142. [CrossRef]

35. Fu, Z.W.; Yu, Y.H.; Li, Z.; Xiao, M.; Han, D.M.; Wang, S.J.; Meng, Y.Z. Surface reduced CeO_2 nanowires for direct conversion of CO_2 and methanol to dimethyl carbonate: Catalytic performance and role of oxygen vacancy. *Catalysts* **2018**, *8*, 164. [CrossRef]

36. Chen, Y.; Xiao, M.; Wang, S.J.; Han, D.M.; Lu, Y.X.; Meng, Y.Z. Porous diatomite-immobilized Cu–Ni bimetallic nanocatalysts for direct synthesis of dimethyl carbonate. *J. Nanomater.* **2012**, 1–8. [CrossRef]

37. Bonzel, H.P.; Bradshaw, A.M.; Ertl, G. *Physics and Chemistry of Alkali Metal Adsorption*; Elsevier: Amsterdam, The Netherlands, 1989.

38. Kazi, A.M.; Chen, B.; Goodwin, J.G.; Marcelin, G.; Rodriguez, N.; Baker, T.K. Li^+ promotion of Pd/SiO_2: The effect on hydrogenation, hydrogenolysis, and methanol synthesis. *J. Catal.* **1995**, *157*, 1–13. [CrossRef]

39. Evin, H.N.; Jacobs, G.; Ruiz-Martinez, J.; Thomas, G.A.; Davis, B.H. Low temperaturewater-gas shift: Alkali doping to facilitate formate C-H bond cleaving over Pt/ceria catalysts—An optimization problem. *Catal. Lett.* **2008**, *120*, 166–178. [CrossRef]

40. Pigos, J.M.; Brooks, C.J.; Jacobs, G.; Davis, B.H. Low temperature water-gas shift: The effect of alkali doping on the C-H bond of formate over Pt/ZrO_2 catalysts. *Appl. Catal. A Gen.* **2007**, *328*, 14–26. [CrossRef]

41. Graetsch, H.; Gies, H.; Topalovic, I. NMR, XRD and IR study on microcrystalline opals. *Phys. Chem. Miner.* **1994**, *21*, 166–175. [CrossRef]

catalysts

MDPI

Article

Metal-Carbon-CNF Composites Obtained by Catalytic Pyrolysis of Urban Plastic Residues as Electro-Catalysts for the Reduction of CO_2

Jesica Castelo-Quibén, Abdelhakim Elmouwahidi, Francisco J. Maldonado-Hódar, Francisco Carrasco-Marín and Agustín F. Pérez-Cadenas *

Carbon Materials Research Group, Department of Inorganic Chemistry, Faculty of Sciences, University of Granada, Avenida de Fuentenueva, s/n, Granada ES18071, Spain; jesicacastelo@ugr.es (J.C.-Q.); aelmouwahidi@ugr.es (A.E.); fjmaldon@ugr.es (F.J.M.-H.); fmarin@ugr.es (F.C.-M.)
* Correspondence: afperez@ugr.es; Tel.: +34-958-24-33-16

Received: 26 March 2018; Accepted: 7 May 2018; Published: 9 May 2018

Abstract: Metal–carbon–carbon nanofibers composites obtained by catalytic pyrolysis of urban plastic residues have been prepared using Fe, Co or Ni as pyrolitic catalysts. The composite materials have been fully characterized from a textural and chemical point of view. The proportion of carbon nanofibers and the final content of carbon phases depend on the used pyrolitic metal with Ni being the most active pyrolitic catalysts. The composites show the electro-catalyst activity in the CO_2 reduction to hydrocarbons, favoring all the formation of C1 to C4 hydrocarbons. The tendency of this activity is in accordance with the apparent faradaic efficiencies and the linear sweep voltammetries. The cobalt-based composite shows high selectivity to C3 hydrocarbons within this group of compounds.

Keywords: carbon dioxide; CO_2 electro-reduction; metal-carbon-CNF composites; plastic waste; carbon-based electrodes; carbon nanofibers

---

1. Introduction

The increase of CO_2 concentration in the atmosphere is thought to be one of the main causes of global climate change [1]. In particular, the CO_2 emission from the use of fossil fuels contributes to the increasing concentration because it establishes a continuous net increase in the natural cycle of the tropospheric carbon.

There are different strategies proposed to control this issue [2] being the most extended the CO_2 storage, and the CO_2 transformation to other valuable products, together with the implementation of renewable energies.

On the other hand, renewable energy sources are supposed to be a replacement, but nowadays they are not producing the constant currents that fossil fuels provide. For this reason, the storage of surplus electrical energy produced during the peak production periods, and its release during peak demand periods, should be crucial. In this manner, extensive research effort has focused on battery storage [3]. However, battery manufacturing requires a lot of resources, reducing their contribution to controlling CO_2 emission, and its life is relatively limited. Furthermore, recycling of their components is also a challenge.

One possible option to address the problem of temporary storing and local surplus of renewable energy is the electro-catalytic reduction of CO_2 to hydrocarbons in water [4]. In this process, the water is split to provide the required hydrogen atoms, which react with CO_2 to form hydrocarbons that can be used directly in the existing infrastructure of fuel transportation as well as in storing the renewable energy.

The direct electrochemical reduction of CO_2 in aqueous solution has been typically studied with metal electrodes like Cu, Au, or Sn during the past few decades [5–10]. Copper electrodes have been found to be quite good in the reduction of CO_2 to hydrocarbons, although the Faradaic efficiency was still low as a result of the dissociation of H_2O to H_2 [9]. More recently, metallic electrodes derived from corresponding metal oxides, like SnO_x, seemed to show promising results in certain catalytic performance for CO_2 reduction [11–13], and only a few transition-metal oxides such as TiO_2, FeO_x and Cu_2O have been reported as potential electro-catalysts for this application [14].

Alternatively, the application of carbon materials in electro-catalytic CO_2 reduction process is a plausible option, which has been tested with platinum catalysts supported on carbon nanotubes, carbon cloth or carbon black [15,16], and even metal-free carbons [17]. Centi et al. [15] showed for the first time the possibility of electro-catalytically converting CO_2 to hydrocarbons with carbon chains >C5, and with product distributions which do not follow the Anderson–Schulz–Flory distribution model typical for Fischer-Tropsch synthesis. Li et al. [17] addressed nanoporous S-doped and S,N-codoped carbons as catalysts for electrochemical reduction CO_2 to CO and CH_4, where the negative charge on the pyridinic nitrogen groups promotes electron–proton transfer to CO_2 leading to COOH* intermediates, which are further reduced to CO.

Carbon gels doped with transition metals have also shown activity in this reaction [18,19]. Although this CO_2 reduction mechanism is still being studied [20], the products obtained in the direct electrochemical reduction of CO_2 to hydrocarbons can achieve several carbon atoms [21]. Regarding the hydrocarbon selectivity, recently [22,23], a high selectivity to C3-hydrocarbons among the detected products has been reported using Co- and Fe-carbon electrodes. Moreover, Fe-carbon electrodes have shown a well-fitted linear correlation between the average crystal sizes of iron and the faradaic efficiencies: the smaller the crystal size, the higher the faradaic efficiency [23].

On the other hand, the large amounts of plastic residue is also a very important environmental problem, and the out-of-control combustion method should not be an option, because it would increase the CO_2 atmospheric and cause other environmental pollution problems [24]. Within these materials, polyethylene (HDPE or LDPE) based plastic bags represent a significant proportion. There are several propositions for the recycling of plastic waste, as their transformation in fuels [25], the recovering of valuable components [26] of their transformation in carbon-based materials or composites [27]. However, the direct application of carbon-metal composites, obtained from real world plastic waste as CO_2 electro-catalysts have not been reported yet. One important advantage of these composite materials as electro-catalysts compared to other proposed carbon based electro-catalysts is the low cost of the raw material since they can be obtained directly from the plastic waste. Therefore, with this proposal, we are focusing our actions on the CO_2 problem twice, (i) researching in its electro-catalytic transformation to hydrocarbons; and (ii) proposing a way for the transformation of LDPE based residues in valuable products.

In the present work, we demonstrate the application of metal-carbon-carbon nanofibers composites obtained from real world plastic waste as promising electrodes in the electro-catalytic reduction of CO_2 to hydrocarbons.

2. Results and Discussion

2.1. Textural and Chemical Characterization of the Composites

Table 1 summarizes the surface areas and pore volumes of the composites. These materials show apparent surface areas between 44 and 80 $m^2 \ g^{-1}$, and an extremely low microporosity, in fact, their N_2 adsorption isotherms show typical type IV shapes.

SEM images of the composites are collected in Figure 1. The morphology consists in a mixture of carbon particles pseudo-flats, overlapped among them from where carbon-nanofibers (CNF) emerge. These CNF are clearly visible in sample PNi, where they are longer than in the other composites, although by HRTEM (Figure 2) the presence of CNF has been detected in all the samples being these CNF clearly hollow and maintaining the metal particle inside of them in most cases. Moreover, the main metal particles Fe, Co or Ni are very well dispersed throughout the carbon matrix (Figure 2), showing a wide range of sizes but all within the nanometric scale.

Table 1. Name, surface area and pore volumes of the composites.

Sample	S_{BET} ($m^2 g^{-1}$)	W_0 ($cm^3 g^{-1}$)	L_0 (nm)	$V_{0.95}$ ($cm^3 g^{-1}$)
PFe	44	0.01	2.08	0.100
PCo	80	0.02	2.16	0.189
PNi	44	0.01	2.18	0.149

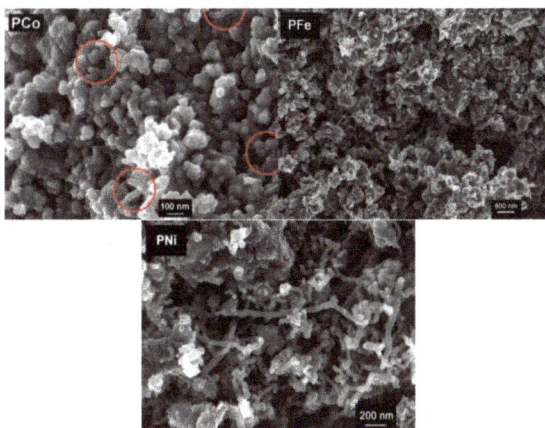

Figure 1. SEM microphotographs showing the morphology of the different composites. Red circles indicate CNF in sample PCo.

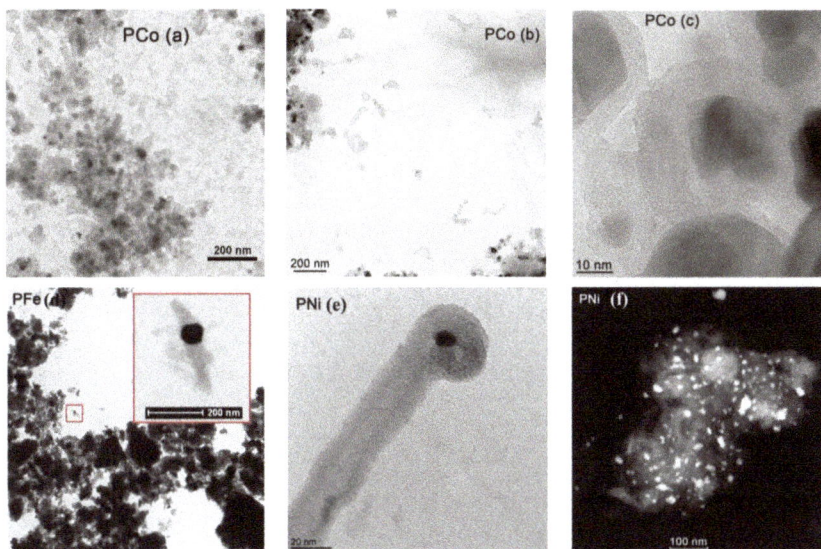

Figure 2. (a–e) High-resolution transmission electron microscopy (HRTEM) images of the composites; (f) scanning transmission electron microscopy (STEM) image, using a high angle annular dark field (HAADF) detector of the PNi sample.

Regarding the metal chemical composition of the composites, Table 2 collects the ICP analysis of the samples. Firstly, it is observed that the content of the pyrolitic catalyst is not the same among the composites which denote a different catalytic behavior and, therefore, indicates different yield in carbon phases. In this line, composite PFe shows the highest metal loading, followed by PCo and lastly by PNi. Thus, small amounts of Fe, Co or Ni are present in all the samples which are in agreement with other studies where the metal content of commercial plastic bags have been analyzed [28]. Much more significant are the relatively high contents of Ca and K; these metals are typical additives of the commercial plastic bags to enhance stiffness and mechanical properties [29].

Table 2. Metal content of the composites determined by inductively coupled plasma optical emission spectrometry (ICP-OES).

Sample	Ca_{ICP} (wt. %)	Co_{ICP} (wt. %)	Fe_{ICP} (wt. %)	K_{ICP} (wt. %)	Ni_{ICP} (wt. %)
PCo	2.31	9.93	0.36	2.56	0.12
PFe	3.75	0.07	15.92	4.68	0.43
PNi	1.56	0.02	0.29	2.02	6.52

On the other site, Fe, Co, and Ni nanoparticles are mainly reduced in zero oxidation state (Figure 3) and situated in both of the carbon phases. This conclusion results from the corresponding XRD peaks that are clearly visible at 44.7°, 44.4° and 44.5° for Fe, Co, and Ni, respectively, while practically negligible amounts of these metals were detected by XPS (Table 3). Moreover, the aforementioned Fe, Co or Ni nanoparticles situated in the amorphous carbon phase are covered by graphitic clusters as it has been observed by HRTEM (Figure 2c), which is due to the fact that these metals are also the catalytic metals of graphitization [30,31].

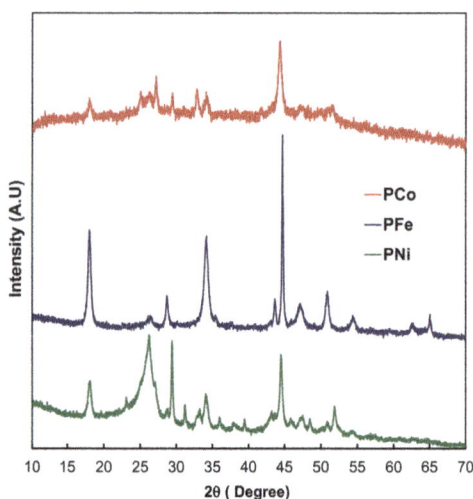

Figure 3. XRD patterns of the composites.

Table 3. Superficial chemical content of the composites determined by XPS.

Sample	C (wt. %)	O (wt. %)	K (wt. %)	Ca (wt. %)	Co (wt. %)	Fe (wt. %)	Ni (wt. %)
PCo	87.71	6.93	2.05	2.97	0.34	0.00	0.00
PFe	82.21	8.53	5.35	3.81	0.00	0.10	0.00
PNi	89.84	4.12	3.91	1.98	0.00	0.00	0.15

Nevertheless, the case of Ca and K are very different, also between them: Ca particles are clearly Ca(II) forming part of Ca(OH)$_2$; XRD peaks at 18 and 34° seem to confirm it (Figure 3) and the Ca2p$_{3/2}$ XPS peak at 348.4 eV as well (Figure 4). However, K particles have not been detected by XRD, which means that its sizes should be lower than 4 nm approx., however, K(0) peaks have been clearly detected by XPS at 293.7 eV (Figure 4; only XPS spectra of PNi has been included in this Figure being the corresponding PCo and PFe XPS spectra very similar to this one). A plausible explanation would be that K(0) nanoclusters or atoms could be mainly inserted between the graphenic layers, as it is only in this situation that they could they keep the zero oxidation state after the final washing treatment of the samples. Finally, it should be noted that both data of chemical analysis, ICP and XPS are in good agreement: PNi is the composite with the lowest transition metal content obtained by ICP analysis and with the highest carbon content determined by XPS, while in the case of PFe, opposites occur. On the other hand, both K and Ca are homogeneously distributed throughout the composite since similar contents are obtained from both techniques.

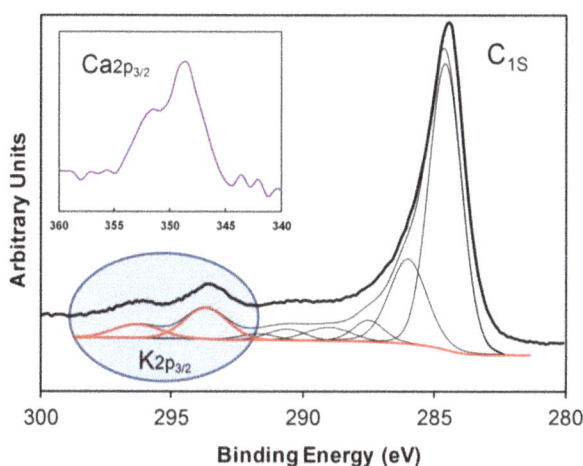

Figure 4. XPS spectra of composite PNi.

2.2. Electro-Reduction of CO$_2$

The three composites were used as a cathode in the electro-catalytic reduction of CO$_2$. A graphite sheet with dimensions of 50 mm × 8 mm was also tested as a blank cathode in the electro-catalytic reduction of CO$_2$. The products analyzed in the gas phase of the reactor were the following: methane (CH$_4$), ethane (C$_2$H$_6$), ethene (C$_2$H$_4$), propane (C$_3$H$_8$), propene (C$_3$H$_6$), propyne (C$_3$H$_4$) and n-butane (C$_4$H$_{10}$); Figure 5 shows the evolution of these compounds vs. the reaction time.

The molar production has also been described regarding C1, C2, C3 and C4 hydrocarbons, 1 to 4 being the number of carbon atoms in the molecules to simplify the discussion about the reaction selectivity. Nevertheless, methane was the major product in all of the cases, and minor amounts of other detected products (probably C4 isomers, or C5, hydrocarbons) have not been quantified. Thus, it is necessary to clarify that when an electrolyte free from dissolved CO$_2$ was used (that is, carrying out the reaction under Ar-saturated solution), hydrocarbons, CO or CO$_2$ were not detected in any cases. Similarly, in the presence of CO$_2$ (normal experimental reaction conditions) and using only the pure graphite cathode, no hydrocarbons were detected. Regarding the molar production, although the surface area and pore volume of the composites were relatively low, the adsorption of a part of the product in the porous structure of the carbon phases could never be ruled out.

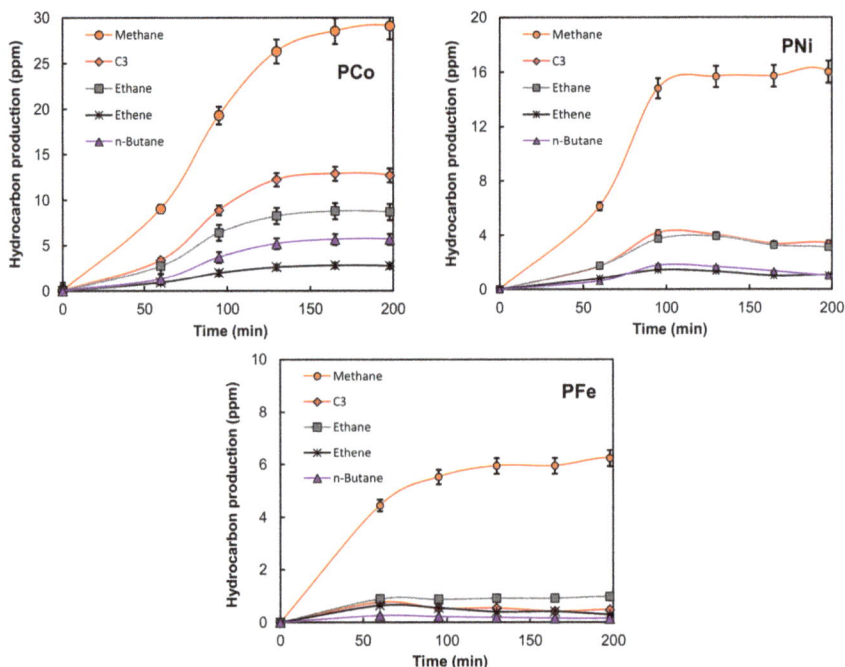

Figure 5. Product distribution (ppm) vs. reaction time.

Data of the total molar productions of hydrocarbons in the reactor are collected in Figure 6; from this Figure, an adequate comparison of the electro-catalytic behavior of these composites is not straightforward because they do not present similar textural characteristics or metal contents. On the other side, we can see that the rate of hydrocarbon formation tends to decrease after ~100 min of reaction for PNi and PFe, and somewhat later for PCo. This type of catalytic behavior has been previously observed [22] and explained by a high formation of H_2 and O_2, which can provoke a dilution effect of the hydrocarbon formation in the gas phase of the reactor [32]. In any case, the following catalytic tendency can clearly be observed: PCo > PNi > PFe, which is in accordance with the apparent faradaic efficiencies calculated at 95 min of reaction time (Table 4). It is important to clarify that that only the detected hydrocarbons (nor H_2 neither O_2 formation) have been included in the faradaic efficiencies calculations. Faradaic efficiencies of this order have been previously obtained with Fe and Co doped carbon gels [22,23].

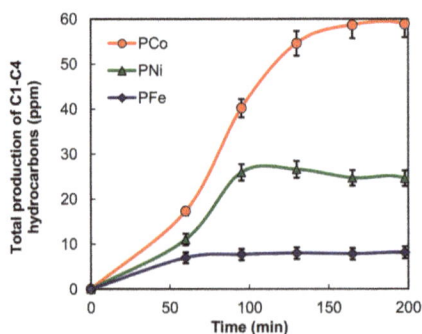

Figure 6. Molar production vs. time obtained with the electro-catalysts.

Table 4. Mean crystal particle size (d_{XRD}), and apparent faradaic efficiencies (F.E.) for the electro-catalytic CO_2 reduction determined at 95 min of reaction time at -1.65 V vs. Ag/AgCl.

Composite	d_{XRD} * (nm)	F.E. (%)
PCo	13.6	0.46
PNi	20.7	0.40
PFe	39.4	0.06

(*) Co, Ni or Fe, respectively.

Therefore, the current findings demonstrate that all composites work as electro-catalysts in this reaction and that they are able to produce the CO_2 transformation to at least C4 hydrocarbons. Moreover, the composite PCo exhibits high selectivity to C3 products (Figure 7) within this group of detected products, which is in accordance with a recent finding with another type of Co doped carbon materials [22]. On the other hand, although PNi is mainly selective to CH_4, the amount of C3 produced was also higher than C2. Finally, composite PFe was the least active catalyst, and it was also least selective to long chain hydrocarbons, despite it having the highest metal loading (Table 2). As previously reported, this behavior is most likely due to its large mean crystal size of Fe (Table 4) [23]. In this line, it should be noted that this composite has a mean metal particle size much larger than the other samples. Figure 8 compares the LSV curves of CO_2 reduction obtained with all the electro-catalysts. The strong increase of the current values at -1.65 V of the composite PFe denotes a low electro-catalytic capacity of this cathode in comparison with the other two samples.

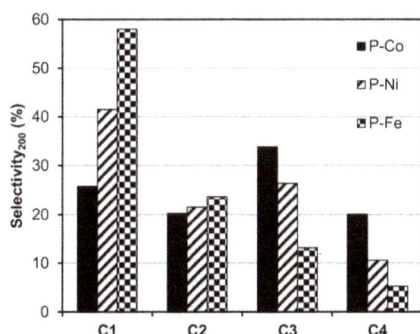

Figure 7. Product distribution (%) in terms of carbon selectivity in gas phase products after 200 min of reaction time.

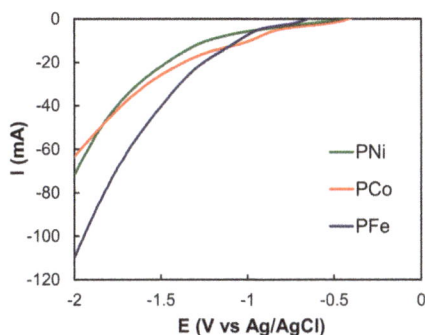

Figure 8. Linear sweep voltammetries obtained from the equilibrium electrode potential to a negative electric potential of -2.00 V vs. Ag/AgCl. Scan rate: 5 mV/s. Fresh electrodes in CO_2 saturated 0.1 M $KHCO_3$.

Finally, all these cathodes were used in a second run of reaction, renovating the electrolyte and saturating with CO_2; all of them again showed similar profiles from those collected in Figure 6. After drying, no significant weight differences were detected between the cathodes after the first and the second runs. Transition metal leaching in the liquid solutions was studied in all cases by ICP-OES; the detected metal concentrations in the liquids always being lower than 100 ppb.

This work shows that the preparation of CO_2 electro-catalyst from real world plastic waste is achievable and although both the preparation method and the electro-chemical application still needs improvement and/or optimization for a real application, the obtained results are very promising. To our knowledge, the direct application of metal-carbon-CNF composites obtained from the real world plastic waste as CO_2 electro-catalysts has not been reported yet.

3. Materials and Methods

Three different composites of metal-carbon-carbon nanofibers (-CNF) were prepared by a catalyzed pyrolysis of urban plastic residues which were thermally pre-treated in a closed reactor. These residues were plastic bags that were used in several well-known supermarkets in Spain, in which the polymer composition mainly consisted of low-density polyethylene (LDPE). Firstly, 10 g of the above-mentioned plastic bags were dissolved in 100 mL of o-xylene at 80 °C, and then 2 g of the catalysts precursor was added. The catalyst precursors were the corresponding hydroxides of Fe, Co and Ni, and the resulting mixture was stirred for 4 h. After that, the o-xylene was evaporated, and the solid was heat treated at 350 °C in a closed reactor (Parr Instrument, Moline, IL, USA) (reactor ref. A1828HC2) for four hours. Finally, the so pre-treated solid was pyrolised under N_2 flow (300 mL min^{-1}) at 900 °C. Before the characterization, the composites were washed with cool water several times. The obtained composites were named as PFe, PCo and PNi, being Fe, Co and Ni the pyrolitic catalyst, respectively.

The metal contents of the composites were determined by inductively coupled plasma optical emission spectrometry (ICP-OES) using an ICP-OES PerkinElmer OPTIMA 8300 spectrometer (PerkinElmer, Madrid, Spain).

The samples were texturally characterized by physical adsorption of nitrogen, scanning electron microscopy (SEM), high resolution transmission electron microscopy (HRTEM), and chemically characterized by X-ray diffraction (XRD) and X-ray photoelectron spectroscopy (XPS) Linear sweep voltammetries (LSV) were also carried out.

N_2 adsorption was carried out at −196 °C. Prior to this measuring process, the samples were outgassed overnight at 110 °C under high vacuum (10^{-6} mbar). The BET equation was applied to the N_2 adsorption data obtaining the apparent surface area, S_{BET}. The Dubinin-Radushkevich (DR) equation was applied to the N_2 adsorption data to obtain the corresponding micropore volume (W_0) and micropore mean width (L_0). Total pore volumes ($V_{0.95}$) were calculated from N_2 adsorption isotherms at −196 °C and at 0.95 relative pressure.

SEM was carried out using a Zeiss SUPRA40VP scanning electron microscope (Carl Zeiss AG, Oberkochen, Germany), equipped with a secondary electron detector, back-scatter electron detector and by using an X-Max 50 mm energy dispersive X-ray microanalysis system. All the samples were crushed before performing this analysis.

HRTEM was performed using a FEI Titan G2 60–300 microscope (FEI, Eindhoven, The Netherlands) with a high brightness electron gun (X-FEG) operated at 300 kV and equipped with a Cs image corrector (CEOS), and for analytical electron microscopy (AEM) a SUPER-X silicon-drift window-less EDX detector.

XRD analysis was carried out BRUKER D8 ADVANCE diffractometer (BRUKER, Rivas-Vaciamadrid, Spain) using CuKα radiation. JCPDS files were searched to assign the different diffraction lines observed. Diffraction patterns were recorded between 10° and 70° (2θ) with a step of 0.02° and a time per step of 96 s. The average crystal size was determined using the Scherrer equation.

XPS measurements of the metal-carbon-CNF composites were performed using a Physical Electronics ESCA 5701 (PHI, Chanhassen, MN, USA) equipped with a MgKα X-ray source (hυ = 1253.6 eV) operating at 12 kV and 10 mA and a hemispherical electron analyzer. The obtained binding energy (BE) values were referred to the C_{1s} peak at 284.7 eV. A base pressure of 10^{-9} mbar was maintained during data acquisition. The survey and multi-region spectra were recorded at C_{1s}, O_{1s}, Fe_{2p}, Co_{2p}, Ni_{2p}, K_{2p} and Ca_{2p} photoelectron peaks. Each spectral region was scanned enough times to obtain adequate signal-to-noise ratios. The spectra obtained after the background signal correction were fitted to Lorentzian and Gaussian curves to obtain the number of components, the position of each peak, and the peak areas.

Electro-catalytic reduction of CO_2 to hydrocarbons was carried out in a three-electrode cell, working in batch mode at ambient temperature and pressure. The cell has 300 cm^3 of total capacity. A Biologic VMP multichannel potentiostat (Bio-Logic Spain, Barcelona, Spain) was used to induce and control the electro-catalytic reaction by applying the selected potential differences over the electrodes. A platinum electrode was used as a counter electrode and Ag/AgCl as a reference electrode. The used electrolyte was 150 cm^3 of CO_2-saturated 0.1 M potassium bicarbonate aqueous solution. The setup was used in potentiostatic mode at −1.65 V, reproducing the voltage conditions of previous works [19]. Prior to the electro-catalytic CO_2 reduction, the liquid phase was saturated through bubbling with CO_2 for 3 h. After saturation, the pH of the solution was 6.7. The CO_2 feed and exit lines were closed off and the reactor was operated in the batch mode. The amount of composite used in the cathode as electro-catalyst (working electrode) was 80 mg which was homogeneously pasted on both faces of a graphite sheet with dimensions of 50 mm × 8 mm. In the preparation of the cathode, the metal-carbon-CNF composite was mixed with the corresponding amount of polytetrafluoroethylene (PTFE) in a weight ratio of (80:7) using a PTFE (60%) water solution. All working electrodes were kept in 0.1 M potassium bicarbonate aqueous solution overnight before being used in the electro-reactor. The samples were also tested as electro-catalysts carrying out the reaction under Ar-saturated solution and, therefore, using electrolytes free of CO_2.

The samples were also characterized by LSV (Bio-Logic Spain, Barcelona, Spain). The cathodic sweep analysis was conducted from the equilibrium electrode potential to negative electric potential of −2.0 V vs. Ag/AgCl, with a scan rate of 5 mV s^{-1}, using the same experimental conditions and reactor set-up for the electro-catalytic reduction of CO_2.

The hydrocarbons produced by the electro-chemical reduction of CO_2 were analyzed from the gas phase using a gas chromatograph (GC) (Bruker Española, Rivas-Vaciamadrid, Spain), where the gases were directly injected into the GC column using a gas recirculating pump for low flows. The GC (carrier gas: He, column: Chrompack Poraplot Q, 50 m × 0.53 mm) was equipped with a FID and TCD detectors. The distribution of gaseous products can be expressed in terms of the carbon selectivity as the amount of carbon atoms (from CO_2) in a specific product relative to the total amount of carbon atoms in the detected hydrocarbons.

$$S_{C_i}(\%) = \frac{i \cdot n_{C_i}}{\sum_i i \cdot n_{C_i}} \times 100\%$$

Here n_{C_i} represents the mol of product C_i, and i the number of carbon atoms in that product.

The liquid phase was also analyzed by Headspace Gas Chromatography-Mass Spectrometry using another GC equipped with a HP-INNOWax 30 m × 0.25 mm × 0.25 μm column, which was coupled to a MS-Triple quadrupole. The presence of carboxylic acids or alcohols of one to four carbon atoms were not detected.

4. Conclusions

Metal-carbon-CNF composites have been obtained from the urban plastic waste. The amount and type of carbon nanofibers and final carbon contents depend on the pyrolitic used metal: Fe, Co or Ni.

Ni catalysts yield the major amount of CNF and carbon phases in the composites. On the other hand, significant contents of Ca and K are also present in the composites, however, while Ca is forming part of $Ca(OH)_2$, K atoms could be embedded inside the carbon phases as K metallic. The composites have been tested as electro-catalyst in the CO_2 reduction to hydrocarbons, and all of them promoted the formation of C1 to C4 hydrocarbons with different activity: PCo > PNi > PFe, which is in accordance with the apparent faradaic efficiencies. It should be highlighted that PCo shows high selectivity to C3 products within this group of compounds.

Author Contributions: A.F.P.-C. and F.C.-M. conceived and designed the experiments; J.C.-Q. and A.E. performed the experiments; A.F.P.-C., F.C.-M. and F.J.M.-H. analyzed the data; A.F.P.-C. and F.C.-M. wrote the paper.

Acknowledgments: This research is supported by the FEDER and Spanish projects CTQ2013-44789-R (MINECO) and P12-RNM-2892 (Junta de Andalucía). J.C.-Q. is grateful to the Junta de Andalucía for her research contract (P12-RNM-2892). A.E. acknowledges a predoctoral fellowship from Erasmus Mundus, Al-Idrissi, programme.

Conflicts of Interest: The authors declare no conflict of interest.

References

1. World Meteorological Organization. Available online: http://www.wmo.int/ (accessed on 1 May 2018).
2. Centi, G.; Perathoner, S. Opportunities and prospects in the chemical recycling of carbon dioxide to fuels. *Catal. Today* **2009**, *148*, 191–205. [CrossRef]
3. Stevenson, K. The origin, development, and future of the lithium-ion battery. *J. Solid State Electrochem.* **2012**, *16*, 2017–2018. [CrossRef]
4. Bevilacqua, M.; Filippi, J.; Miller, H.A.; Vizza, F. Recent technological progress in CO_2 electroreduction to fuels and energy carriers in aqueous environments. *Energy Technol.* **2015**, *3*, 197–210. [CrossRef]
5. Hori, Y.; Murata, A. Electrochemical evidence of intermediate formation of adsorbed CO in cathodic reduction of CO_2 at a nickel electrode. *Electrochim. Acta* **1990**, *35*, 1777–1780. [CrossRef]
6. Hori, Y.; Murata, A.; Takahashi, R. Formation of hydrocarbons in the electrochemical reduction of carbon dioxide at a copper electrode in aqueous solution. *J. Chem. Soc. Faraday Trans.* **1989**, *85*, 2309–2326. [CrossRef]
7. Hori, Y.; Wakebe, H.; Tsukamoto, T.; Koga, O. Electrocatalytic process of CO selectivity in electrochemical reduction of CO_2 at metal electrodes in aqueous media. *Electrochim. Acta* **1994**, *39*, 1833–1839. [CrossRef]
8. Jitaru, M.; Lowy, D.A.; Toma, M.; Toma, B.C.; Oniciu, L. Electrochemical reduction of carbon dioxide on flat metallic cathodes. *J. Appl. Electrochem.* **1997**, *27*, 875–889. [CrossRef]
9. Gattrell, M.; Gupta, N.; Co, A. A review of the aqueous electrochemical reduction of CO_2 to hydrocarbons at copper. *J. Electroanal. Chem.* **2006**, *594*, 1–19. [CrossRef]
10. Chaplin, R.P.S.; Wragg, A.A. Effects of process conditions and electrode material on reaction pathways for carbon dioxide electroreduction with particular reference to formate formation. *J. Appl. Electrochem.* **2003**, *33*, 1107–1123. [CrossRef]
11. Zhang, R.; Lv, W.; Li, G.; Lei, L. Electrochemical reduction of CO_2 on SnO_2/nitrogen-doped multiwalled carbon nanotubes composites in $KHCO_3$ aqueous solution. *Mater. Lett.* **2015**, *141*, 63–66. [CrossRef]
12. Li, F.; Chen, L.; Knowles, G.P.; MacFarlane, D.R.; Zhang, J. Hierarchical Mesoporous SnO_2 Nanosheets on Carbon Cloth: A Robust and Flexible Electrocatalyst for CO_2 Reduction with High Efficiency and Selectivity. *Angew. Chem. Int. Ed.* **2017**, *56*, 505–509. [CrossRef] [PubMed]
13. Bashir, S.; Hossain, S.; Rahman, S.u.; Ahmed, S.; Amir, A.; Hossain, M.M. Electrocatalytic reduction of carbon dioxide on SnO_2/MWCNT in aqueous electrolyte solution. *J. CO_2 Util.* **2016**, *16*, 346–353. [CrossRef]
14. Zhu, D.D.; Liu, J.L.; Qiao, S.Z. Recent Advances in Inorganic Heterogeneous Electrocatalysts for Reduction of Carbon Dioxide. *Adv. Mater.* **2016**, *28*, 3423–3452. [CrossRef] [PubMed]
15. Centi, G.; Perathoner, S.; Wine, G.; Gangeri, M. Electrocatalytic conversion of CO_2 to long carbon-chain hydrocarbons. *Green Chem.* **2007**, *9*, 671–678. [CrossRef]
16. Centi, G.; Perathoner, S. Problems and perspectives in nanostructured carbon-based electrodes for clean and sustainable energy. *Catal. Today* **2010**, *150*, 151–162. [CrossRef]
17. Li, W.; Seredych, M.; Rodriguez-Castellon, E.; Bandosz, T.J. Metal-free nanoporous carbon as a catalyst for electrochemical reduction of CO_2 to CO and CH_4. *ChemSusChem* **2016**, *9*, 606–616. [CrossRef] [PubMed]

18. Perez-Cadenas, A.F. Doped Carbon Material for the Electrocatalytic Conversion of CO_2 into Hydrocarbons, Uses of the Material and Conversion Method Using Said Material. Patent WO/2013/004882, 1 October 2013.

19. Perez-Cadenas, A.F.; Ros, C.H.; Morales-Torres, S.; Perez-Cadenas, M.; Kooyman, P.J.; Moreno-Castilla, C.; Kapteijn, F. Metal-doped carbon xerogels for the electro-catalytic conversion of CO_2 to hydrocarbons. *Carbon* **2013**, *56*, 324–331. [CrossRef]

20. Schouten, K.J.P.; Kwon, Y.; van der Ham, C.J.M.; Qin, Z.; Koper, M.T.M. A new mechanism for the selectivity to C1 and C2 species in the electrochemical reduction of carbon dioxide on copper electrodes. *Chem. Sci.* **2011**, *2*, 1902–1909. [CrossRef]

21. Qiao, J.; Liu, Y.; Hong, F.; Zhang, J. A review of catalysts for the electroreduction of carbon dioxide to produce low-carbon fuels. *Chem. Soc. Rev.* **2014**, *43*, 631–675. [CrossRef] [PubMed]

22. Abdelwahab, A.; Castelo-Quibén, J.; Pérez-Cadenas, M.; Elmouwahidi, A.; Maldonado-Hódar, F.J.; Carrasco-Marín, F.; Pérez-Cadenas, A.F. Cobalt-doped carbon gels as electro-catalysts for the reduction of CO_2 to hydrocarbons. *Catalysts* **2017**, *7*, 25. [CrossRef]

23. Castelo-Quiben, J.; Abdelwahab, A.; Perez-Cadenas, M.; Morales-Torres, S.; Maldonado-Hodar, F.J.; Carrasco-Marin, F.; Perez-Cadenas, A.F. Carbon-iron electro-catalysts for CO_2 reduction. The role of the iron particle size. *J. CO_2 Util.* **2018**, *24*, 240–249. [CrossRef]

24. Keane, M.A. Catalytic Transformation of Waste Polymers to Fuel Oil. *ChemSusChem* **2009**, *2*, 207–214. [CrossRef] [PubMed]

25. Aguado, J.; Serrano, D.P.; San Miguel, G.; Castro, M.C.; Madrid, S. Feedstock recycling of polyethylene in a two-step thermo-catalytic reaction system. *J. Anal. Appl. Pyrol.* **2007**, *79*, 415–423. [CrossRef]

26. Quicker, P. Thermal Treatment as a Chance for Material Recovery. In *Source Separation and Recycling: Implementation and Benefits for a Circular Economy*; Maletz, R., Dornack, C., Ziyang, L., Eds.; Springer International Publishing: Cham, Switzerland, 2018; pp. 119–149.

27. Bazargan, A.; McKay, G. A review—Synthesis of carbon nanotubes from plastic wastes. *Chem. Eng. J.* **2012**, *195*, 377–391. [CrossRef]

28. Huerta-Pujol, O.; Soliva, M.; Giró, F.; López, M. Heavy metal content in rubbish bags used for separate collection of biowaste. *Waste Manag.* **2010**, *30*, 1450–1456. [CrossRef] [PubMed]

29. Yang, K.; Yang, Q.; Li, G.; Sun, Y.; Feng, D. Morphology and mechanical properties of polypropylene/calcium carbonate nanocomposites. *Mater. Lett.* **2006**, *60*, 805–809. [CrossRef]

30. Maldonado-Hodar, F.J.; Moreno-Castilla, C.; Rivera-Utrilla, J.; Hanzawa, Y.; Yamada, Y. Catalytic graphitization of carbon aerogels by transition metals. *Langmuir* **2000**, *16*, 4367–4373. [CrossRef]

31. Maldonado-Hodar, F.J.; Moreno-Castilla, C.; Perez-Cadenas, A.F. Surface morphology, metal dispersion, and pore texture of transition metal-doped monolithic carbon aerogels and steam-activated derivatives. *Microporous Mesoporous Mater.* **2004**, *69*, 119–125. [CrossRef]

32. Goncalves, M.R.; Gomes, A.; Condeco, J.; Fernandes, R.; Pardal, T.; Sequeira, C.A.C.; Branco, J.B. Selective electrochemical conversion of CO_2 to C2 hydrocarbons. *Energy Convers. Manag.* **2010**, *51*, 30–32. [CrossRef]

catalysts

MDPI

Article

Surface Oxidation of Supported Ni Particles and Its Impact on the Catalytic Performance during Dynamically Operated Methanation of CO$_2$

Benjamin Mutz [1,2], Andreas Martin Gänzler [1], Maarten Nachtegaal [3], Oliver Müller [4,5], Ronald Frahm [4], Wolfgang Kleist [1,2,6] and Jan-Dierk Grunwaldt [1,2,*]

[1] Institute for Chemical Technology and Polymer Chemistry, Karlsruhe Institute of Technology (KIT), D-76131 Karlsruhe, Germany; benjamin.mutz@kit.edu (B.M.); andreas.gaenzler@kit.edu (A.M.G.); wolfgang.kleist@rub.de (W.K.)

[2] Institute of Catalysis Research and Technology, Karlsruhe Institute of Technology (KIT), D-76344 Eggenstein-Leopoldshafen, Germany

[3] Paul Scherrer Institute (PSI), CH-5232 Villigen, Switzerland; maarten.nachtegaal@psi.ch

[4] Department of Physics, University of Wuppertal, D-42119 Wuppertal, Germany; omueller@slac.stanford.edu (O.M.); frahm@uni-wuppertal.de (R.F.)

[5] Stanford Synchrotron Radiation Lightsource (SSRL), SLAC National Accelerator Laboratory, Menlo Park, CA 94025, USA

[6] Laboratory of Industrial Chemistry, Ruhr-University Bochum, D-44801 Bochum, Germany

* Correspondence: grunwaldt@kit.edu; Tel.: +49-721-608-42120

Received: 1 August 2017; Accepted: 5 September 2017; Published: 18 September 2017

Abstract: The methanation of CO$_2$ within the power-to-gas concept was investigated under fluctuating reaction conditions to gather detailed insight into the structural dynamics of the catalyst. A 10 wt % Ni/Al$_2$O$_3$ catalyst with uniform 3.7 nm metal particles and a dispersion of 21% suitable to investigate structural changes also in a surface-sensitive way was prepared and characterized in detail. *Operando* quick-scanning X-ray absorption spectroscopy (XAS/QEXAFS) studies were performed to analyze the influence of 30 s and 300 s H$_2$ interruptions during the methanation of CO$_2$ in the presence of O$_2$ impurities (technical CO$_2$). These conditions represent the fluctuating supply of H$_2$ from renewable energies for the decentralized methanation. Short-term H$_2$ interruptions led to oxidation of the most reactive low-coordinated metallic Ni sites, which could not be re-reduced fully during the subsequent methanation cycle and accordingly caused deactivation. Detailed evaluation of the extended X-ray absorption fine structure (EXAFS) spectra showed surface oxidation/reduction processes, whereas the core of the Ni particles remained reduced. The 300-s H$_2$ interruptions resulted in bulk oxidation already after the first cycle and a more pronounced deactivation. These results clearly show the importance and opportunities of investigating the structural dynamics of catalysts to identify their mechanism, especially in power-to-chemicals processes using renewable H$_2$.

Keywords: CO$_2$ methanation; dynamic reaction conditions; *operando* XAS; quick-EXAFS; surface oxidation-reduction; H$_2$ dropout

1. Introduction

The power-to-chemicals concept is an important strategy for future renewable energy systems based on chemical energy storage. The splitting of water to produce H$_2$ and the catalytic conversion of CO$_2$ to methane, alkanes, methanol or higher alcohols are the main steps to generate a chemical energy carrier [1–4]. Both steps need to withstand fluctuations in supplied electricity from wind and solar plants, which occur temporarily and fluctuate on a time scale of seconds to days. When using a small or even no H_2 reservoir, these fluctuations are transferred to the catalytic reactor imposing a varying supply of H_2 used to hydrogenate CO_2 [5–8].

Methane (synthetic or substitute natural gas) as one possible chemical energy carrier can be distributed easily and unlimitedly in the existing natural gas grid. The dynamics in the methanation originating from fast load changes of the H_2 supply are a key challenge [5] and are therefore the subject of present research. The catalytic methanation of CO_2 is well-known, and numerous articles can be found in the literature reporting CO_2 conversion and CH_4 selectivity under optimized and steady state conditions [9,10]. Ni-based catalysts generally achieve good activity. In addition, due to the low cost and high activity, Ni catalysts have emerged as the most commonly-used methanation catalysts [11–18], since more active Ru catalysts [19,20] suffer from the high price of the noble metal. The most relevant catalyst support material in industrial methanation reactions is reported to be γ-Al_2O_3 with a high surface area [19,21,22].

Methanation under transient reaction conditions has been performed to gain a better understanding of the mechanism, investigating the adsorbed species while switching between different gas mixtures [23,24]. Simulations of dynamic methanation reactors have been conducted to gain kinetic data and to address problems such as overheating during the transient process [25,26]. Other systems that have been studied under transient reaction conditions are, e.g., the Fischer–Tropsch reaction, where the influence of unsteady state conditions on the process itself was investigated [7,8,27]. Various spectroscopic, microscopic and diffraction techniques applied under reaction conditions revealed different structural changes of catalysts used in processes operated under changing reaction atmospheres [28–33].

Previous studies have shown that during the dynamic methanation of technical CO_2, a fast partial oxidation of Ni-based catalysts occurred in a less reducing atmosphere after a H_2 dropout, which caused a deactivation in the following methanation sequence due to the presence of inactive NiO on the metallic catalyst [22]. Further deactivation occurred over cycles, but the initial catalytic activity was recovered using reactivation in H_2 at elevated temperatures [34]. In general, interruptions in the H_2 feed must be prevented in methanation processes using CO_2 with traces of oxygen (technical CO_2), since otherwise reactivation is required to retrieve high catalytic performance. Hence, methanation reactors have to be kept under a reducing atmosphere during stand-by operation [6]. Within this study, we provide a more detailed insight into the kinetics and mechanisms of the redox processes in methanation applications using CO_2 directly produced from biogas plants or exhaust gas from power plants containing traces of oxygen [35,36]. Based on these considerations, highly time-resolved *operando* XAS (quick-EXAFS (QEXAFS)) experiments were performed to evaluate the sensitivity of the catalyst towards deactivation. Furthermore, experiments including fast switches of the gas atmospheres were performed to obtain insights into blocking of active sites under fast load changes and thus mechanistic aspects in the methanation of CO_2.

2. Results and Discussion

2.1. Preparation, Characterization and Catalytic Performance

A catalyst featuring uniform and rather small Ni particles was essential to determine the structural changes both of the surface and the bulk of the Ni particles during dynamically operated methanation. The homogeneous deposition-precipitation method was chosen to obtain a Ni/Al_2O_3 catalyst with these properties. Elemental analysis using optical emission spectroscopy with an inductively-coupled plasma (ICP-OES) confirmed a catalyst loading of 10 wt % Ni/Al_2O_3. The specific surface area of the catalyst was 200 m^2/g, and the mean pore diameter was 11 nm. X-ray diffraction (XRD) was not suitable to analyze the catalyst, since the Ni and NiO reflections were either superimposed by the signals of the support, the particles were too small to create reflections or the phases were X-ray amorphous, respectively (cf. the Supplementary Materials, XRD patterns in Figure S1).

As shown in Figure 1, electron microscopy revealed a homogeneous dispersion of the particles on the support resulting in small Ni particles with a diameter of 3.7 ± 1.2 nm, a narrow size distribution and a Ni dispersion of 21%. These values are comparable with those reported in the literature for supported Ni catalysts that were prepared using the same preparation technique [37,38].

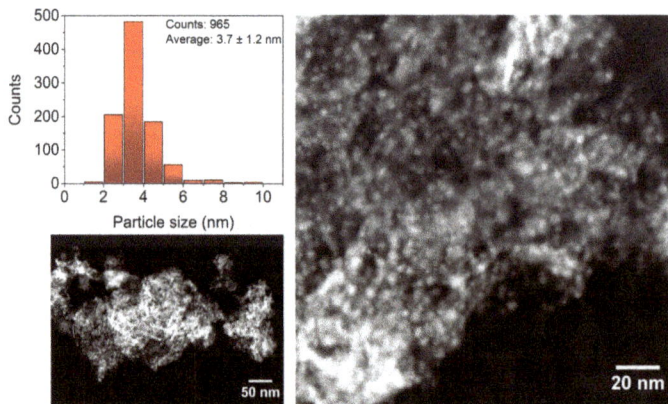

Figure 1. Scanning transmission electron microscopy (STEM) images and particle size distribution (top left) of the 10 wt % Ni/Al$_2$O$_3$ catalyst prepared by homogeneous deposition-precipitation.

As an additional characterization tool, temperature programmed reduction (TPR) experiments were used, which showed H$_2$ consumption in a small temperature range (peak maximum at 565 °C, cf. Figure S2), confirming the uniform particle size and the absence of agglomerates or larger particles. Furthermore, the Ni particles were reduced at lower temperature compared to other Ni/Al$_2$O$_3$ catalysts reported in the literature (peak maxima between 680 and 800 °C [39–41]), which suggests a weaker metal-support interaction in the catalyst prepared in this study due to the synthesis method.

The catalyst was applied in the methanation of CO$_2$ and showed high conversion and selectivity with CO being the only by-product (Figure 2). As the carbon balance was almost closed and therefore the calculation of the conversion according to Equations (1) and (2) (cf. Section 3) was similar, the formation of other products in significant amounts can be excluded. In Figure 2, the values obtained from Equation (1) are shown.

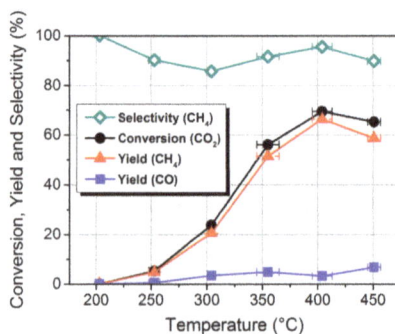

Figure 2. Conversion of CO$_2$ (black), yield of CH$_4$ (red) and yield of CO (blue), as well as selectivity of CH$_4$ (green) for the 10 wt % Ni/Al$_2$O$_3$ catalyst; conditions: stainless steel tubular fixed-bed reactor, 150 mg catalyst, H$_2$/CO$_2$ = 4, 75% N$_2$, p = 1 atm, T = 200–450 °C, gas hourly space velocity (GHSV) of 26,700 h^{-1} and weight hourly space velocity (WHSV) of 12,000 mL$_{CO_2}$/(g$_{cat}$·h).

The formation of CH_4 started between 200 and 250 °C, and the conversion of CO_2 increased with rising temperature, reaching the highest value of 69% at 400 °C at a maximum selectivity towards CH_4 of 95%. At 450 °C, the conversion of CO_2 declined, and the selectivity shifted more toward CO due to the endothermic back reaction and the reverse water-gas shift reaction, which are preferred at higher temperatures [42]. The results are comparable with reports in the literature using Ni/Al_2O_3 catalysts treated under related reaction conditions [19,43]. The turnover frequency (TOF) at 250 °C was calculated as 0.02 s^{-1} or as 0.08 s^{-1} at 300 °C. The TOFs for Ni/Al_2O_3 methanation catalysts reported in the literature cover a broad range between 0.69 s^{-1} [11], 0.10 s^{-1} [18], 0.041–0.097 s^{-1} [12] and 0.5×10^{-3}–2.4×10^{-3} s^{-1} [17]. The achieved TOFs correspond to typical dimensions; however, it is difficult to compare the results due to the diverse reaction conditions applied in these studies. In conclusion, the catalyst used in this study can be regarded as a representative methanation catalyst, which is well suited for our experiments to evaluate the effects of fluctuating reaction conditions on the structure of the active component and their impact on the catalyst activity.

2.2. Operando QEXAFS Studies under Transient Reaction Conditions

Operando X-ray absorption spectroscopic (XAS) studies were performed to gather information on the oxidation state and the atomic structure of nickel during methanation conditions and, especially, during short periods of hydrogen dropout. In previous studies, we observed a significant oxidation of nickel particles, when H_2 was withdrawn from the reaction feed, leading to a lower catalytic activity during the subsequent methanation cycle [22,34]. In those studies, however, the hydrogen feed was interrupted for long periods of time (approximately 1 h) since conventional XAS methods were used, which require relatively long acquisition times per spectrum (approximately 5 min). Therefore, in the present study, we applied the quick-scanning EXAFS (QEXAFS) technique [44–46] to obtain data even in the sub-second regime for studying the impact of short hydrogen dropouts (30–300 s).

The methanation of technical CO_2 containing traces of oxygen and the simulation of the H_2 dropouts were performed at 400 °C (highest conversion and selectivity, compare Figure 2) and the same weight hourly space velocity (WHSV) applied in the activity measurements. Before each experiment, the catalyst was reduced in 50% H_2/He at 500 °C. To simulate a short dropout of H_2, caused for example by fluctuations in the hydrogen feed or by an unexpected change of the operation mode, hydrogen was switched on and off. The results of the experiment are presented in Figure 3.

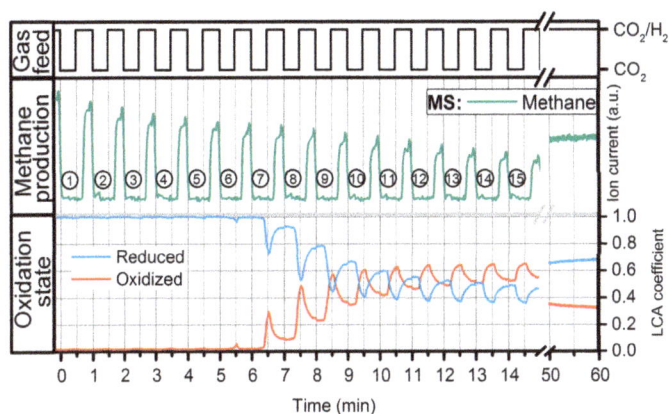

Figure 3. Methanation of CO_2 during dynamic operation, switching every 30 s between methanation conditions ($H_2/CO_2 = 4$) and CO_2 at constant WHSV of 12,000 $mL_{CO_2}/(g_{cat}\cdot h)$ and GHSV of 71,700 h^{-1}. The figure shows the valve signal in the upper part (black), the CH_4 signal of the mass spectrometer (m/z 15) in the middle part (green) and the fraction of reduced (blue) and oxidized (red) Ni from LCA of the X-ray absorption near edge structure (XANES) spectra. The numbers in circles count the H_2 dropouts.

First, the fully-reduced catalyst was exposed to methanation conditions ($H_2/CO_2 = 4$; 75% He). After several minutes of operation and a stable mass spectrometer (MS) signal of methane, the catalyst was subjected to a period of fluctuating H_2 feed for 15 min. During this modulation, the feed was switched every 30 s between 5% CO_2/He and methanation conditions ($H_2/CO_2 = 4$). Finally, the catalyst was again exposed to steady state methanation conditions. The applied feed composition expressed as the signal of the valve position is depicted in the upper part of Figure 3. The product gas was monitored during the entire experiment; all analyzed components are shown in the Supplementary Materials (Figure S14). The methane production was monitored by mass spectrometry (m/z 15), and the resulting signal is presented in the middle part of Figure 3. Throughout the experiment, X-ray absorption spectra were recorded with an acquisition rate of 44 spectra/s. The X-ray absorption near edge structure (XANES, exemplary sequence, see Figure 4) spectra were evaluated with linear combination analysis (LCA, details, cf. the Supplementary Materials) to monitor the Ni oxidation state during transient conditions. The results are plotted in the bottom part of Figure 3.

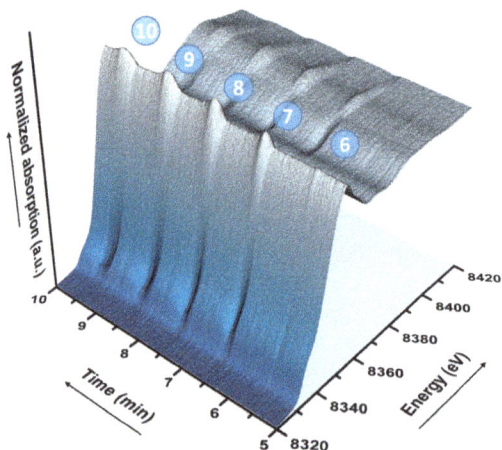

Figure 4. Section of the X-ray absorption near edge structure (XANES) spectra during the 30-s H_2 dropout modulation (dropout Nos. 6–10 according to Figure 3).

The Ni particles were reduced during the methanation of CO_2 ($H_2/CO_2 = 4$), and an intense signal assigned to methane was observed in the mass spectrometer data. Short H_2 cut-offs, however, strongly affected the catalyst and its activity. As H_2 was switched back on the stream after a 30-s period of a less reducing atmosphere without H_2 (0–0.5 min), a slightly decreased methane signal was observed. The methane production decreased further with each H_2 dropout. However, no significant change of the Ni oxidation state was monitored during the first 5 min of the experiment. During the sixth H_2 free period (after 5.5 min), a first slight oxidation of Ni to a fraction of 6% was observed, which was still accounted to be negligible. Fast and significant oxidation of Ni during the seventh H_2 dropout (at 6–6.5 min) was observed by XAS, indicated by the increasing intensity of the white line. As discussed later in this paper, the oxidation of Ni in the absence of H_2 was caused by the traces of oxygen present in the technical CO_2. The oxidation stopped at 29% oxidized Ni according to LCA when H_2 was switched back to the feed at 6.5 min. Even though the catalyst was exposed to the H_2 containing atmosphere for 30 s (at 6.5–7 min), it did not regain its initial state, and 9% of the Ni atoms remained oxidized. The percentage of oxidized Ni species increased steadily during the subsequent modulation until the 12th H_2 dropout, during which approximately 65% of the Ni atoms were oxidized. In the following oxidation and reduction events (between 11 and 15 min), a marginal but no significant increase in the percentage of oxidized Ni was observed, and the fraction of oxidized Ni remained at a constant level altering between 65% and 50% over cycles. Nevertheless, the deactivation of the catalyst continued steadily as shown by the decreasing MS signal of methane.

During the ensuing 10 min of modulation, the changes of the catalyst (altering oxidation/reduction and continuous deactivation) followed the same trend and are therefore not shown here (full range experiment, see Figure S5). After the period of fluctuating operation (25 min in total), a sequence of steady state methanation was performed during which the oxidation state of the catalyst and structure was further monitored. During a 30-min measuring period, significant reduction of Ni was observed over time. This process, however, was very slow, and even after 30 min (final 5 min of this sequence shown in Figure 3), the initial oxidation state was not regained and 30% still remained oxidized. Consequently, the catalyst was less active and showed a similar methane production and oxidized fraction as observed after the ninth H_2 dropout. This supports a correlation between the oxidized fraction of Ni and the formation of methane [22]. Obviously, the temperature of 400 °C was too low to reduce Ni completely, which would have necessitated a re-activation of the catalyst at elevated temperatures [34].

The active catalyst state was obviously very sensitive towards oxidation, and already, a short exposure to a less reducing atmosphere was sufficient to initiate oxidation of Ni resulting in lower activity. Therefore, the full extended X-ray absorption fine structure (EXAFS) spectra were further evaluated to extract information on the local structural changes of the Ni catalyst during fluctuating conditions. In Figure 5a, Fourier-transformed (FT) EXAFS spectra are presented along with best fitting parameters obtained by fitting the first two coordination shells (Figure 5b). The EXAFS evaluation of the cycles 6–8 is shown in Figure 6; the structural parameters and the fully analyzed dataset are listed in Table 1; more details are shown in the Supplementary Materials.

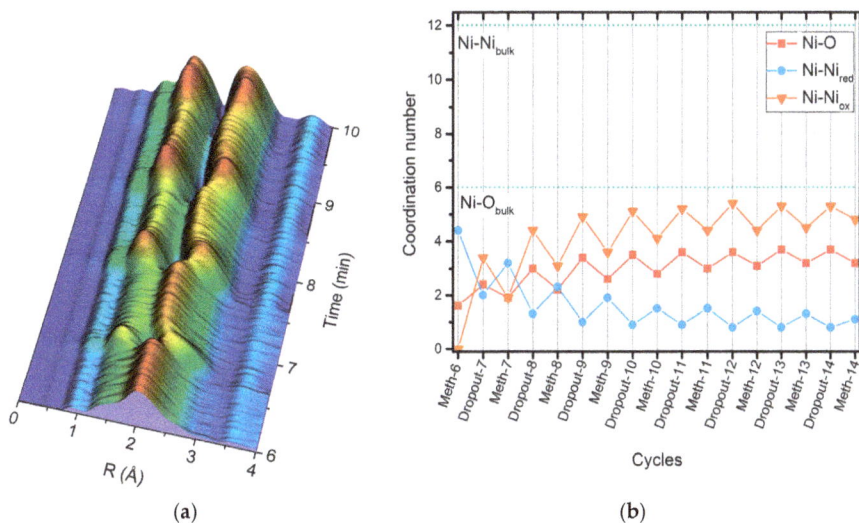

(a)   (b)

Figure 5. (a) Fourier-transformed extended X-ray absorption fine structure (FT EXAFS) data (k: 3–11 Å^{-1}; k^2-weighted, not phase corrected) during 30-s H_2 dropouts (starting at the methanation sequence before the seventh H_2 dropout (Meth-6)); (b) Results of the EXAFS fitting analysis during the 30-s modulation. The final spectra of each sequence (methanation sequences labeled as "Meth") were analyzed (details, cf. the Supplementary Materials). The coordination numbers (N) of neighboring atoms are presented: Ni-O and Ni-Ni$_{ox}$ correspond to O and Ni coordination numbers, respectively, in oxidized nickel and Ni-Ni$_{red}$ to the coordination number of the first nickel shell in reduced nickel.

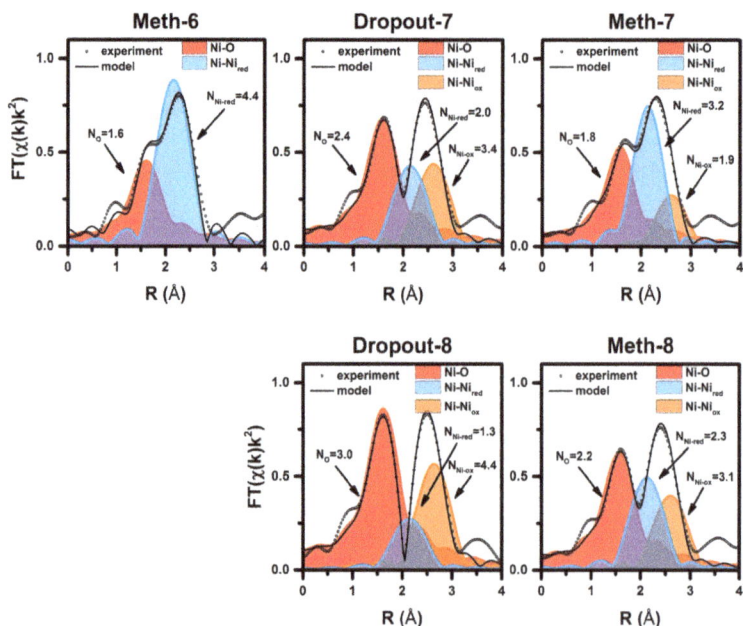

Figure 6. Experimental and simulated FT EXAFS data (magnitude, not phase shift corrected) of the Ni catalyst during 30-s modulation, in addition, the estimated Ni coordination numbers (Cycles 6–8), k-range 3–9 Å^{-1}, k-weight = 2 (for details, see the Supplementary Materials).

Table 1. Structural parameters of the local Ni atomic environment extracted from the best fitting EXAFS spectra shown in Figure 6. k range = 3–9 Å^{-1}, R range = 1.2–3.2 Å, N^{ind} = 6, N^{var} = 5, a = fitted uncertainty 0.01 or lower, f = fixed during fit, N = number of neighboring atoms, R = distance, σ^2 = mean square deviation of interatomic distance, R factor = misfit between experimental data and theory (for details, see the Supplementary Materials).

Cycle	Atom	N	R (Å)^a	$\sigma^2 \cdot 10^{-3}$ (Å^2)	E_0 (eV)	R Factor (%)
Meth-6	O	1.6 ± 0.2	2.03 ± 0.2	8.3^f	9.0 ± 1.5	0.27
	Ni	4.4 ± 0.3	2.49 ± 0.1	11.8^f		
Dropout-7	O	2.4 ± 0.2	2.02 ± 0.03	8.3^f	7.6 ± 2.9	0.25
	Ni	2.0 ± 0.6	2.48 ± 0.02	11.4^f		
	Ni	3.4 ± 0.9	2.96 ± 0.03	11.8^f		
Meth-7	O	1.8 ± 0.2	2.01 ± 0.2	8.3^f	6.4 ± 2.2	0.23
	Ni	3.2 ± 0.4	2.48 ± 0.1	11.4^f		
	Ni	1.9 ± 0.6	2.95 ± 0.3	11.8^f		
Dropout-8	O	3.0 ± 0.2	2.04 ± 0.02	8.3^f	8.9 ± 2.1	0.14
	Ni	1.3 ± 0.6	2.48 ± 0.03	11.4^f		
	Ni	4.4 ± 0.7	2.97 ± 0.03	11.8^f		
Meth-8	O	2.2 ± 0.2	2.02 ± 0.03	8.3^f	6.8 ± 2.9	0.23
	Ni	2.3 ± 0.6	2.48 ± 0.01	11.4^f		
	Ni	3.1 ± 0.9	2.96 ± 0.03	11.8^f		

For the first six cycles, no significant changes in the EXAFS spectra were observed. The spectra were dominated by a feature at about 2 Å, predominantly originating from Ni neighbors in reduced Ni. The evaluated interatomic distance of 2.49 Å was similar to the one in metallic Ni. The O contribution determined in the reduced state may originate from the oxide support or incomplete reduction of the Ni particles. Including a Ni-Ni$_{ox}$ (the Ni-Ni backscattering contribution in nickel oxide) bond pair led to a worse fit and was therefore not considered in Meth-6. During the seventh H$_2$ dropout, when significant oxidation was observed for the first time (cf. LCA in Figure 3), changes in the FT EXAFS data were observed, indicating substantial changes in the structure of the Ni nanoparticles. During oxidation, the feature at 2 Å in the EXAFS spectra resulting from reduced Ni-Ni coordination disappeared quickly and was replaced by two peaks related to the O and Ni neighbors in a NiO-like structure. The evaluation of the final EXAFS spectra of each sequence (methanation atmosphere or less reducing atmosphere, respectively) facilitated monitoring the evolution of the respective coordination shells during fluctuating operation. The respective coordination numbers are therefore plotted in Figure 5b for this purpose. The data revealed the transformation from a reduced Ni particle to a partially-oxidized Ni particle. Even though the Ni particles were very small, they did not get fully oxidized due to the short duration of H$_2$ dropout indicating that a core of reduced Ni remained (coordination number $N_{Ni-Ni,red} > 1$). After the 12th dropout, predominantly a reversible oxidation and reduction was observed, in accordance with the LCA discussed earlier, and the core of the particles remained reduced.

The following steady state methanation operation (full-range experiment, see Figure S5) slowly transformed the catalyst back into its reduced state. However, even after 30 min in the methanation atmosphere, still, a significant fraction of nickel remained oxidized, also shown by the coordination numbers ($N_{Ni-O} = 2.2$, $N_{Ni-Ni,red} = 2.4$, $N_{Ni-Ni,ox} = 3.0$, cf. Table S3). During this methanation experiment, only slight sintering was observed by an increase in the coordination number of reduced Ni by 0.2. This is within the accuracy of the evaluation; however, the Ni coordination number increased by 0.5 after repeating the whole experiment of 30-s modulation three times further (see Table S2). Note that the evaluated Ni coordination was not adjusted to the fraction of the particles in the Ni/NiO mixture, and no conclusion on the correlation between coordination number and particle size was drawn.

The delayed start of the Ni oxidation as observed in Figure 3 was rationalized when the spatial conditions of the experiment were considered [32,47]. XAS spectra were recorded in the middle of the catalyst bed with a 200 μm-sized X-ray beam and correlated to the integral activity data. Catalyst oxidation therefore probably started at the reactor inlet and propagated through the catalyst bed. We speculate that the oxidation was caused by the O$_2$ background present in the CO$_2$ feed (cf. the estimation in the later part of the paper), which oxidized a small amount of Ni at the reactor inlet in the short 30-s period in a less reducing atmosphere and was thereby completely consumed. Significant oxidation by CO$_2$ was ruled out, as CO$_2$ was present in excess and should then also have oxidized Ni in the middle of the reactor already at the beginning of the experiment (cf. Figure 3; oxidation started at the sixth switch). Furthermore, an additional experiment in the total absence of O$_2$ using short and thin stainless steel tubing and a CO$_2$ (99.998%) bottle directly placed next to the mass flow controllers showed that no catalyst deactivation occurred during 60 s H$_2$ dropouts (see Figure S17). Catalyst deactivation occurred during 300-s cycles when 300 ppm and 500 ppm O$_2$ were added to the gas mixture, which was more pronounced with increasing O$_2$ content (Figure S18). Finally, the oxidation of Ni by CO$_2$ is thermodynamically unfavorable ($\Delta G > 0$). Estimations of the amount of O$_2$ present in the spectroscopic experiment based on the formation of NiO showed that an amount of around 1000 ppm was sufficient for the oxidation (for details, cf. the discussion of Figure S7 in the Supplementary Materials). Reduction of the whole catalyst bed was not achieved within the following 30 s during methanation at 400 °C. Therefore, the overall fraction of oxidized Ni increased during fluctuating operation, causing the steady decrease in catalytic activity. The methane production decreased continuously, even after Ni reached a stable oxidized fraction at the monitored position (after the 12th H$_2$ dropout), since Ni oxidation kept propagating towards the end of the catalyst bed.

To provoke a faster oxidation through the catalyst bed and to investigate the oxidation and its kinetics in a longer cycle duration, the catalyst was exposed to H$_2$ dropout events to an extended period of 300 s in a subsequent experiment using the same catalyst sample. In this experiment

(Figure 7), oxidation was observed immediately during the first dropout event. However, there was a significant delay (approximately 80 s; details, see Figure S7) between the switch of the gas atmospheres and the oxidation observed in the XANES spectra, due to the gradual oxidation of the Ni along the catalyst bed. Again, oxidation was found to be fast, resulting in 91% of NiO at the end of the 300-s cycle. Switching to methanation conditions after 300 s led to a re-reduction of the Ni particles and methane formation. About 38% of oxidized nickel remained on the catalyst and changed the overall catalyst composition resulting in a decreased catalytic activity in the second methanation cycle. Reduction occurred approximately 20 s after the gas switch and, therefore, seemed to propagate faster than the oxidation due to the much higher concentration of the reducing agent in the feed. In the following H_2 dropout cycles, the maximum oxidized fraction in the less reducing atmospheres remained at approximately 95%. Therefore, a 300-s H_2 dropout led to a (nearly) complete oxidation of the catalyst. A slowly-progressing reduction of the catalyst was observed during the 30-min period of steady state methanation at the end of the experiment, reaching a final value of a 48% reduced fraction.

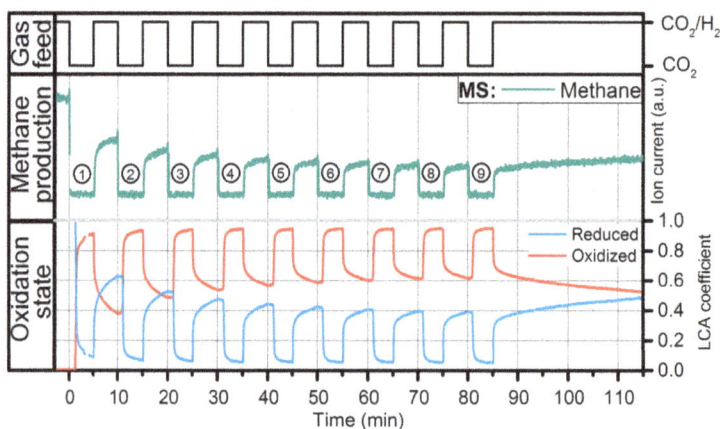

Figure 7. Methanation of CO_2 during dynamic operation switching every 300 s between methanation conditions (H_2/CO_2 = 4) and CO_2 at constant WHSV of 12,000 mL $_{CO_2}$/(g_{cat}·h) and GHSV of 71,700 h^{-1}. The figure shows the valve signal in the upper part (black), the CH_4 signal of the MS (m/z 15) in the middle part (green) and the fraction of reduced (blue) and oxidized (red) Ni from LCA of the XANES spectra. The numbers in cycles count the H_2 dropouts.

Similar to the 30-s modulation, the estimated amount of O_2 based on the formed NiO showed a value around 1000 ppm. Additional experiments with a quantified amount of O_2 (Figure S18) showed a full Ni oxidation after a few 300-s H_2 dropouts applying 500 ppm O_2. Consequently, we assumed an O_2 content between 500 and 1000 ppm present in the QEXAFS experiments. This small quantity is ideal to oxidize the most reactive sites and block in fact those Ni sites for methanation.

The FT EXAFS spectra collected during the 300-s modulation and the coordination numbers from the best fits of the EXAFS spectra are shown in Figure 8a,b, respectively. As already observed during the 30-s modulation, the FT EXAFS spectra in the reduced state were dominated by one signal at 2 Å, which was assigned to Ni-Ni coordination in reduced Ni. Note that the $N_{Ni-Ni,red}$ of 5.1 was slightly higher than the one determined from the EXAFS evaluation during the 30-s modulation ($N_{Ni-Ni,red}$ = 4.4). This can be explained by sintering of the Ni particles since the same sample was used for all experiments and three additional 30-s modulation experiments (not shown here) were performed before the 300-s modulation experiment (see Table S2). Applying the 300-s modulation, the maximum of the Ni-O path was already reached during the first H_2 dropout. A bulk oxidation of the Ni particles occurred during the absence of H_2 indicated by the $N_{Ni-Ni,red}$, which decreased to zero.

The oxidation state reached its steady state after the first cycle, while the amount of receded Ni during the methanation periods decreased until the fourth cycle.

Reduced Ni particles with a $N_{Ni-Ni,red}$ of 5.1 were transformed into oxidized Ni particles with a $N_{Ni-Ni,ox}$ of 7.6. This can be either explained by incomplete reduction of the Ni particles contributing to the apparently lower coordination in reduced Ni and a higher coordination after oxidation. Another explanation for the difference in the coordination numbers can be a wetting effect of the flat reduced particles, increasing the contact with the oxidic support material, whereas the oxidized particles are more spherical with a higher average Ni-Ni coordination [30].

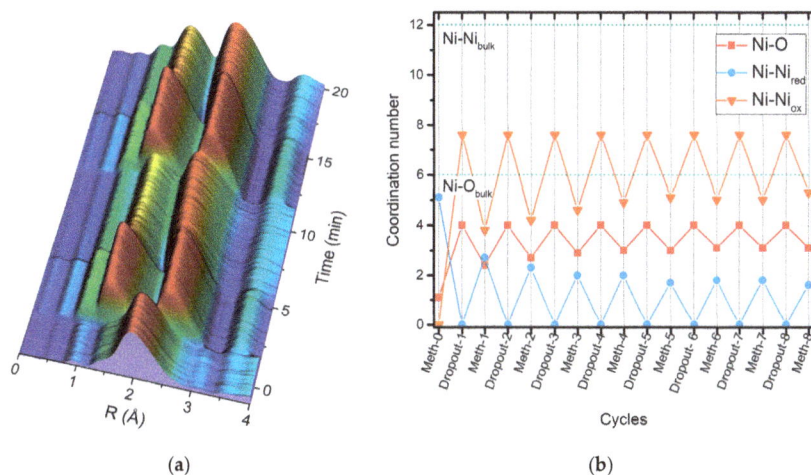

(a)   **(b)**

Figure 8. (a) FT EXAFS data (k: 3–11 Å^{-1}; k^2-weighted) during repeated 300-s H_2 dropouts; (b) Results of the EXAFS fitting analysis during the 300-s modulation. The final spectra of each sequence were analyzed. The coordination numbers of neighboring atoms are presented: Ni-O and Ni-O-Ni correspond to O and Ni coordination numbers, respectively, in oxidized nickel and Ni-Ni to the coordination number of the first nickel shell in reduced nickel.

The direct comparison of the MS signals of both modulation experiments (Figure S16) showed that the deactivation was more distinctive during 300-s switches, where a significantly lower CH_4 signal was detected after the second cycle up until the end of the experiment. Furthermore, in the experiments with 30-s modulation, the catalyst was much more active during steady state methanation after the H_2 dropout sequence. This observation clearly emphasized the correlation of the Ni oxidation state and the methanation activity. Comparing the results of the LCAs from both experiments, full oxidation was interrupted during the 30-s cycles due to the fast switching. During the 300-s modulation, the oxidation state reached a plateau indicating the complete oxidation process. Thus, the impact of the short cycles on the catalyst deactivation was less severe.

Figures 9 and 10 show schemes of the Ni particles during the dynamic methanation of CO_2. Short-term H_2 dropout (Figure 9) first led to an oxidation of the most reactive sites such as low coordinated Ni atoms located at step sites, defect sites or at the interface to the support material [48–52]. Particularly, the interface sites appear important for the methanation of CO_2, since the key steps of the hydrogenation are speculated to take place at the interface of the metal particle and support [53,54]. Furthermore, step sites are more reactive in CO dissociation compared to plane facets [51,55]. NiO islands may form and grow to NiO structures of several layers [56,57]. Diffusion of oxygen into the bulk did not occur due to the short period of less reducing atmosphere. Full oxidation of the Ni particles was interrupted, and NiO planes were reduced partly in the methanation atmosphere.

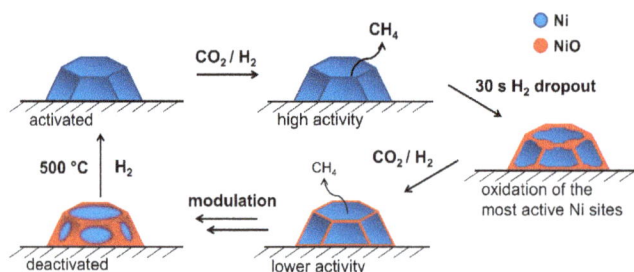

Figure 9. Scheme of the structural changes of supported Ni particles during the 30-s dropout modulations. Blue represents reduced Ni, whereas red describes oxidized Ni.

The highly active lower coordinated Ni sites remained oxidized under these reaction conditions due to their higher stability. Hence, the methanation of CO_2 could only take place at the reduced Ni plane sites resulting in lower activity. This supports earlier studies in the literature [48–50,58] that selective blockage of the low coordinated Ni edge or corner sites leads to differed or deteriorated catalytic properties. Similar structure sensitivity due to easier oxidation of low-coordinated sites was also observed during CO oxidation on Pt catalysts [59,60]. For example, oxidation of low coordinated Pt atoms located at edge and corner sites during CO oxidation was found by XAS resulting in an irreversible CO adsorption on PtO [61]. Cluster models show that metal particles of our size or dispersion contain 4% edge sites [62–65]. This value correlated with the remaining oxidized fraction (10%) in the methanation sequence after the first oxidation during the seventh H_2 dropout when additional sites at the interface, as well as corner and defect sites, would have been taken into account.

The schematic structural changes of Ni particles during 300-s modulations are illustrated in Figure 10. Compared to the short-time H_2 dropouts, the longer 300 s H_2 dropouts resulted in a less reducing atmosphere enabling a complete oxidation of the Ni particles, which then suffer from insufficient re-reduction during the reducing methanation atmosphere.

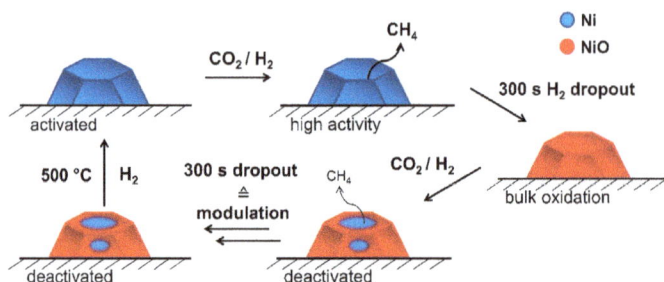

Figure 10. Scheme of the structure of the supported Ni particles exposed to H_2 dropout modulations at a duration of 300 s.

3. Materials and Methods

3.1. Catalyst Preparation

A 10 wt % Ni/Al_2O_3 catalyst was prepared via homogeneous deposition-precipitation using urea as the precipitating agent [66–68]. Al_2O_3 support (Alfa Aesar, Karlsruhe, Germany, 1/8 inch pellets, crushed to fine powder, calcined at 600 °C (5 K/min) for 4 h) was suspended in a solution of 0.03 mol/L $Ni(NO_3)_2 \cdot 6H_2O$ (Merck, Darmstadt, Germany, ≥99%) in water. Five equivalents of urea (Carl Roth, Karlsruhe, Germany, ≥99.6%) were added to the mixture and stirred for 1 h at room temperature, where a pH value of 5 was observed. The suspension was then heated to 90 °C and stirred under reflux for 18 h, cooled to room temperature and stirred for another hour, resulting in a final pH value of 7–8. The solid was filtered, washed with deionized water, dried over night at 110 °C and calcined for 4 h at 500 °C (5 K/min). Typically, 2 g of the catalyst were prepared in one batch containing 1.80 g Al_2O_3, 1.00 g Ni nitrate, 1.04 g urea and 115 mL water.

3.2. Characterization

The chemical composition was determined via optical emission spectroscopy with an inductively-coupled plasma (ICP-OES) using an Agilent (Waldbronn, Germany) 720/725 emission spectrometer. The specific surface area of the catalyst was determined using N_2 physisorption at −196 °C according to the Brunauer–Emmet–Teller (BET) method via multipoint measurements using a BELSORP-mini II (MicrotracBEL, Osaka, Japan). X-ray diffraction (XRD) patterns were recorded in the range 2 θ = 20–80° (step size 0.017°, 0.51 s per scan step) using a rotating sample holder containing catalyst powder and a PANalytical (Almelo, The Netherlands) X'pert PRO diffractometer with Ni-filtered Cu-Kα radiation (1.54060 Å). Scanning transmission electron microscopy (STEM) images were recorded using a FEI (Hillsboro, OR, USA) Titan 80-300 microscope operated at 300 kV and a Fischione Model 3000 high-angle annular dark-field (HAADF-STEM) detector. The measurements were conducted using reduced catalyst powder dispersed on a gold grid covered with holey carbon film. The Ni particle size was determined with ImageJ software (1.48v, National Institutes of Health, Rockville, MD, USA, 2016) measuring the mean diameter of 965 marked particles using ellipsoid shapes. The Ni dispersion was calculated using the mean diameter of the Ni particles, the area occupied by a Ni surface atom and the volume occupied by a Ni atom in the bulk metal assuming spherical particles [69]. A Micromeritics (Aachen, Germany) AutoChem II 2920 chemisorption analyzer was used for temperature programmed reduction (TPR) experiments. One hundred milligrams of the catalyst sample (100–200 µm) were placed in a U-shaped quartz tube and fixed with quartz wool plugs. First, the catalyst was heated to 500 °C (10 K/min) in 10% O_2 followed by the TPR between 40 and 900 °C (10 K/min) using 10% H_2. The H_2 consumption was recorded using a thermal conductivity detector (TCD).

3.3. Catalytic Performance

The catalytic activity in the methanation of CO_2 was tested in a continuous flow laboratory setup with a stainless steel reactor filled with a fixed-bed of 150 mg catalyst sample (300–450 µm, diluted in SiC) operated in down-flow. The reactor was heated by a custom-made oven (HTM Reetz, Berlin, Germany), and the temperature was controlled using a thermocouple placed in front of the catalyst bed inside of the reactor and regulated by a Eurotherm 2416 PID controller. Gas dosing was adjusted using individual mass flow controllers (Bronkhorst, Ruurlo, The Netherlands). First, the catalyst was reduced in situ in 50% H_2/N_2 (600 mL/min) at 500 °C (10 K/min) for 2 h. Then, the reactor was cooled to 200 °C, and the gas atmosphere was switched to H_2/CO_2 = 4 in 75% N_2 (600 mL/min total flow), resulting in a gas hourly space velocity (GHSV) of 26,700 h^{-1} (concerning total gas flow and catalyst volume) or a weight hourly space velocity (WHSV) of 12,000 mL$_{CO_2}$/(g$_{cat}$·h) (concerning gas flow of CO_2 and catalyst mass), respectively. The methanation reaction was performed between 200 and 450 °C in 50 °C steps to determine the optimum reaction conditions. Reactant and product gas compositions were analyzed by a Thermo Scientific (Waltham, MA, USA) C2V-200 micro gas chromatograph (microGC) equipped with a molecular sieve (5 Å) and a QS-BOND column and a thermal conductivity detector (TCD). Conversion, yield and selectivity were calculated as follows:

$$\text{Conversion: } X(CO_2) = \left(1 - \frac{CO_{2,out}}{CO_{2,out} + CH_{4,out} + CO_{out}}\right) \times 100\% \qquad (1)$$

or including N_2 as the internal standard:Conversion:

$$\text{Conversion: } X'(CO_2) = \left(1 - \frac{CO_{2,out} \times N_{2,in}}{N_{2,out} \times CO_{2,in}}\right) \times 100\% \tag{2}$$

$$\text{Yield: } Y(CH_4 \text{ or } CO) = \frac{CH_{4,out} \text{ or } CO_{out}}{CO_{2,out} + CH_{4,out} + CO_{out}} \times 100\% \tag{3}$$

$$\text{Selectivity: } S(CH_4) = \frac{Y(CH_4)}{X(CO_2)} \times 100\% \tag{4}$$

The turnover frequency (TOF) was calculated as moles of CH_4 produced per moles of Ni surface atoms per second using the inlet flow of CO_2, the molar gas volume $V(m)$ and the catalyst mass:

$$\text{TOF} = \frac{\dot{V}(CO_2, in) \times Y(CH_4)}{V(m) \times N(Ni, surf) \times m(cat)} \tag{5}$$

The number of *Ni* surface atoms per gram catalyst $N(Ni, surf)$ was determined using the dispersion from TEM analysis [69].

3.4. Operando XAS Experiments

Quick-scanning X-ray absorption spectroscopy (QEXAFS) [70,71] was performed at the SuperXAS beamline at the Swiss Light Source (SLS) synchrotron facility (Paul Scherrer Institute, Villigen, Switzerland), which operates in top-up mode at 400 mA and 2.4 GeV. The measurements were performed at the Ni K-edge (8333 eV) in transmission mode. The polychromatic X-ray beam was collimated using a Si-coated mirror at 2.5 mrad located in front of the monochromator. The monochromatized beam was focused to a beam size of 200 µm × 200 µm at the sample position using a Rh-coated mirror at 2.5 mrad. A channel-cut Si(111) monochromator oscillating at 22 Hz was used together with N_2-filled gridded ionization chambers both dedicated for fast data acquisition, where a Ni foil was measured simultaneously with the data for absolute energy calibration. In each oscillation of the monochromator, two spectra were recorded, one with increasing energy, another one with decreasing energy. With this system *operando* X-ray absorption spectroscopy (XAS) data were recorded with 44 spectra/s.

For the *operando* XAS experiments the catalyst sample was diluted with Al_2O_3 (1:1) and placed into a 1.5-mm quartz glass capillary (100–200 µm sieved fraction, 10 mg, 10 mm catalyst bed) and fixed with quartz wool plugs [72]. Spectra were acquired at the middle of the catalyst bed. The capillary-based reactor was connected to the gas dosing system equipped with individual mass flow controllers (Bronkhorst, Ruurlo, The Netherlands) and heated using a hot air blower (FMB Oxford GSB-1300, Oxford, UK) [73]. The total gas flow was adjusted to 20 mL/min using similar compositions as in the laboratory experiments (reduction: 50% H_2/He, methanation: 25% reactants (H_2/CO_2 = 4) in He). The same WHSV of 12,000 mL_{CO_2}/(g_{cat}·h) as in the laboratory was used resulting in a higher GHSV of 71,700 h^{-1} due to the different dimensions of the reactor. Fast switches of the gas atmosphere were realized using a micro-electric VICI (Schenkon, Switzerland) 4-way valve. The gas composition was analyzed by a Pfeiffer Vacuum (Aßlar, Germany) ThermoStar™ GSD 320 mass spectrometer (MS) equipped with a quartz capillary and a C-SEM/Faraday detector.

The measurement of 44 spectra/s allowed detection of changes with a time resolution of 23 ms. However, it was found that a 1-s time resolution was sufficient to monitor the structural changes of the catalyst, and thus, the spectra could be averaged for 1 s each to improve data quality. Matlab® (v8.6, The Mathworks Inc., Natick, MA, USA, 2015) routines were used for merging QEXAFS data to 1 s/spectrum, energy calibration to a metallic Ni foil, normalization and for conducting linear combination analysis (LCA). For the latter, a linear combination of reference spectra was fitted to the sample spectra in the XANES region (from −30 eV to +50 eV). The spectrum of the reduced catalyst in 50% H_2/He atmosphere and of the oxidized catalyst in 5% O_2, both acquired at 400 °C, were used as reduced and oxidized reference, respectively (see Figure S3).

EXAFS data analysis was performed using Athena and Artemis of the IFEFFIT software package (v1.2.11, Chicago, IL, USA, 2008) [74] on spectra collected during steady state conditions and on selected spectra during fluctuating operation. In this case, 220 spectra were averaged, as this was sufficient to monitor the structural changes and improve the data quality (5 s/spectrum). The theoretical data was adjusted to fit the experimental spectra by a least square method in Fourier-transformed (FT) EXAFS spectra (with k^1-, k^2- and k^3-weighted EXAFS functions, FT EXAFS spectra, k and R range, as well as fixed and varied parameters are shown in the Supplementary Materials) considering metallic Ni (ICSD: 64989) and NiO (ICSD: 9866). Corresponding backscattering amplitudes and phases were calculated with Feff 6.0 [75]. Amplitude reduction factors (S_0^2) were obtained on metallic and NiO bulk references (more details, cf. the Supplementary Materials). Coordination number (N), bond distance (R) and energy alignment between theoretical and experimental data (ΔE_0) were refined for each spectrum. The mean square deviation of interatomic distances (σ^2) was obtained by simultaneously refining spectra of the sample in similar conditions and at the same temperature (cf. the Supplementary Materials).

4. Conclusions

Experiments with periodically changing atmospheres during the methanation of technical CO_2 simulated as H_2 dropouts were performed and resulted in detailed insights into the catalyst deactivation mechanism. During H_2 interruptions, catalyst deactivation occurred due to oxidation of the most active surface Ni sites in the presence of O_2 impurities (500–1000 ppm) in technical CO_2. Thirty-second short-term H_2 dropouts showed surface oxidation and reduction occurring after the seventh modulation cycle monitored in the center of the catalyst bed. The core of the Ni particles remained reduced, since full oxidation was interrupted. Nevertheless, catalyst deactivation occurred already after the first dropout since the oxidation of the active sites propagates stepwise through the catalyst bed starting from the reactor inlet. This gives important insight into the mechanism of the methanation reaction and, obviously, even short interruptions in the H_2 feed, and thus, slight oxidation of the most active Ni sites cannot be tolerated by the catalyst. Therefore, it is important to prevent H_2 dropouts in industrial methanation applications, where traces of O_2 might be present. This can also be transferred to other power-to-chemicals processes using renewable H_2. A higher impact on the deactivation was observed during 300-s modulation, where a bulk oxidation was observed even in the first H_2 dropout, which resulted in a higher fraction of oxidized Ni remaining on the catalyst during the methanation sequence and, thus, an even lower CH_4 production.

The presence of small amounts of oxygen in the CO_2 stream can already block a small fraction of most active Ni sites, which had a strong influence on the CO_2-methanation activity. This "titration" of active sites may be further used in future to understand the methanation of CO_2, but also other CO_2 hydrogenation mechanisms in more detail.

Supplementary Materials: The following are available online at www.mdpi.com/2073-4344/7/9/279/s1, Figure S1: XRD patterns of the γ-Al_2O_3 support and the 10 wt % Ni/Al_2O_3 catalyst in its oxidized and reduced state. Figure S2: H_2-TPR profile of the 10 wt % Ni/Al_2O_3 catalyst. Figure S3: Reference XANES spectra of the bulk materials (recorded at room temperature) and in situ oxidized and reduced spectra of the catalyst (measured at 400 °C) for linear combination analysis (LCA). Figure S4: XANES and FT EXAFS spectra of the reduced catalyst in H_2/He at room temperature (RT) and at 400 °C and in He atmosphere at 400 °C. Figure S5: Full-length experiment of the 30-s H_2 dropout modulations during methanation of CO_2. Figure S6: Zoom of the first 6 min during the 30-s modulation. Figure S7: Zoom of the first 14 min during the 300-s H_2 dropout modulation. Figures S8–S13: Experimental and simulated EXAFS data. Figures S14 and S15: Full MS dataset recorded during the 30-s and 300-s H_2 dropout modulations, respectively. Figure S16: Comparison of the MS signals of CH_4 (m/z 15) during the 30-s and 300-s H_2 dropout modulations. Figure S17: Sixty-second H_2 modulation in an oxygen-free experiment. Figure S18: Series of 300-s H_2 dropout periods during the methanation of CO_2 containing 300–500 ppm O_2. Tables S1–S5: Structural parameters of the local Ni atomic environment extracted from the EXAFS spectra.

Acknowledgments: This work was funded and supported by Karlsruhe Institute of Technology (KIT) and the Helmholtz Research School "Energy-Related Catalysis". The Swiss Light Source (SLS) is acknowledged for providing beam time at the SuperXAS beamline and technical support. The infrastructure was funded within the BMBF projects "MatAkt" (05K10VKB), "ZeitKatMat" (05K13VK13, 05K13PX1) and MatDynamics (05K16PX1). The authors thank Di Wang and Sina Baier (STEM), Hermann Köhler (ICP-OES) and Angela Beilmann (BET).

Author Contributions: J.-D.G. and W.K. developed the topic with B.M. and A.M.G and supervised the experimental work. B.M. designed and characterized the catalyst and performed activity measurements. B.M. and A.M.G. performed the *operando* QEXAFS experiment. O.M. and M.N. contributed with technical and scientific support. O.M., R.F. and M.N. developed the QEXAFS setting at SLS. A.M.G. analyzed the EXAFS data supported by O.M. The manuscript was written with contributions from all authors.

Conflicts of Interest: The authors declare no conflict of interest.

References

1. Schüth, F. Chemical Compounds for Energy Storage. *Chem. Ing. Technol.* **2011**, *83*, 1984–1993. [CrossRef]
2. Schlögl, R. *Chemical Energy Storage*; Walter de Gruyter GmbH: Berlin, Germany; Boston, MA, USA, 2013.
3. Sterner, M. *Erneuerbare Energien und Energieeffizienz—Renewable Energies and Energy Efficiency*; Kassel University Press GmbH: Kassel, Germany, 2009; Volume 14.
4. Hashimoto, K.; Yamasaki, M.; Fujimura, K.; Matsui, T.; Izumiya, K.; Komori, M.; El-Moneim, A.A.; Akiyama, E.; Habazaki, H.; Kumagai, N.; et al. Global CO_2 recycling—Novel materials and prospect for prevention of global warming and abundant energy supply. *Mater. Sci. Eng. A* **1999**, *267*, 200–206. [CrossRef]
5. Kalz, K.F.; Kraehnert, R.; Dvoyashkin, M.; Dittmeyer, R.; Gläser, R.; Krewer, U.; Reuter, K.; Grunwaldt, J.-D. Future Challenges in Heterogeneous Catalysis: Understanding Catalysts under Dynamic Reaction Conditions. *Chem. Cat. Chem.* **2017**, *9*, 17–29. [CrossRef] [PubMed]
6. Götz, M.; Lefebvre, J.; Mörs, F.; McDaniel Koch, A.; Graf, F.; Bajohr, S.; Reimert, R.; Kolb, T. Renewable Power-to-Gas: A technological and economic review. *Renew. Energy* **2016**, *85*, 1371–1390. [CrossRef]
7. Iglesias González, M.; Schaub, G. Fischer-Tropsch Synthesis with H_2/CO_2—Catalyst Behavior under Transient Conditions. *Chem. Ing. Technol.* **2015**, *87*, 848–854. [CrossRef]
8. Iglesias González, M.; Eilers, H.; Schaub, G. Flexible Operation of Fixed-Bed Reactors for a Catalytic Fuel Synthesis–CO_2 Hydrogenation as Example Reaction. *Energy Technol.* **2016**, *4*, 90–103. [CrossRef]
9. Gao, J.; Liu, Q.; Gu, F.; Liu, B.; Zhong, Z.; Su, F. Recent advances in methanation catalysts for the production of synthetic natural gas. *RSC Adv.* **2015**, *5*, 22759–22776. [CrossRef]
10. Aziz, M.A.A.; Jalil, A.A.; Triwahyono, S.; Ahmad, A. CO_2 methanation over heterogeneous catalysts: Recent progress and future prospects. *Green Chem.* **2015**, *17*, 2647–2663. [CrossRef]
11. Aziz, M.A.A.; Jalil, A.A.; Triwahyono, S.; Mukti, R.R.; Taufiq-Yap, Y.H.; Sazegar, M.R. Highly active Ni-promoted mesostructured silica nanoparticles for CO_2 methanation. *Appl. Catal. B* **2014**, *147*, 359–368. [CrossRef]
12. Bian, L.; Zhang, L.; Xia, R.; Li, Z. Enhanced low-temperature CO_2 methanation activity on plasma-prepared Ni-based catalyst. *J. Nat. Gas Sci. Eng.* **2015**, *27*, 1189–1194. [CrossRef]
13. Liu, J.; Li, C.; Wang, F.; He, S.; Chen, H.; Zhao, Y.; Wei, M.; Evans, D.G.; Duan, X. Enhanced low-temperature activity of CO_2 methanation over highly-dispersed Ni/TiO_2 catalyst. *Catal. Sci. Technol.* **2013**, *3*, 2627–2633. [CrossRef]
14. Ocampo, F.; Louis, B.; Roger, A.-C. Methanation of carbon dioxide over nickel-based $Ce_{0.72}Zr_{0.28}O_2$ mixed oxide catalysts prepared by sol-gel method. *Appl. Catal. A* **2009**, *369*, 90–96. [CrossRef]
15. Rahmani, S.; Rezaei, M.; Meshkani, F. Preparation of highly active nickel catalysts supported on mesoporous nanocrystalline γ-Al_2O_3 for CO_2 methanation. *J. Ind. Eng. Chem.* **2014**, *20*, 1346–1352. [CrossRef]
16. Tada, S.; Shimizu, T.; Kameyama, H.; Haneda, T.; Kikuchi, R. Ni/CeO_2 catalysts with high CO_2 methanation activity and high CH_4 selectivity at low temperatures. *Int. J. Hydrogen Energy* **2012**, *37*, 5527–5531. [CrossRef]
17. He, S.; Li, C.; Chen, H.; Su, D.; Zhang, B.; Cao, X.; Wang, B.; Wei, M.; Evans, D.G.; Duan, X. A Surface Defect-Promoted Ni Nanocatalyst with Simultaneously Enhanced Activity and Stability. *Chem. Mater.* **2013**, *25*, 1040–1046. [CrossRef]
18. Li, Y.; Zhang, Q.; Chai, R.; Zhao, G.; Liu, Y.; Lu, Y.; Cao, F. Ni-Al_2O_3/Ni-foam catalyst with enhanced heat transfer for hydrogenation of CO_2 to methane. *AlChE J.* **2015**, *61*, 4323–4331. [CrossRef]

19. Garbarino, G.; Bellotti, D.; Riani, P.; Magistri, L.; Busca, G. Methanation of carbon dioxide on Ru/Al_2O_3 and Ni/Al_2O_3 catalysts at atmospheric pressure: Catalysts activation, behaviour and stability. *Int. J. Hydrogen Energy* **2015**, *40*, 9171–9182. [CrossRef]

20. Abe, T.; Tanizawa, M.; Watanabe, K.; Taguchi, A. CO_2 methanation property of Ru nanoparticle-loaded TiO_2 prepared by a polygonal barrel-sputtering method. *Energy Environ. Sci.* **2009**, *2*, 315–321. [CrossRef]

21. Hoekman, S.K.; Broch, A.; Robbins, C.; Purcell, R. CO_2 recycling by reaction with renewably-generated hydrogen. *Int. J. Greenh. Gas Control* **2010**, *4*, 44–50. [CrossRef]

22. Mutz, B.; Carvalho, H.W.P.; Mangold, S.; Kleist, W.; Grunwaldt, J.-D. Methanation of CO_2: Structural response of a Ni-based catalyst under fluctuating reaction conditions unraveled by *operando* spectroscopy. *J. Catal.* **2015**, *327*, 48–53. [CrossRef]

23. Aldana, P.A.U.; Ocampo, F.; Kobl, K.; Louis, B.; Thibault-Starzyk, F.; Daturi, M.; Bazin, P.; Thomas, S.; Roger, A.C. Catalytic CO_2 valorization into CH_4 on Ni-based ceria-zirconia. Reaction mechanism by *operando* IR spectroscopy. *Catal. Today* **2013**, *215*, 201–207. [CrossRef]

24. Zarfl, J.; Ferri, D.; Schildhauer, T.J.; Wambach, J.; Wokaun, A. DRIFTS study of a commercial Ni/γ-Al_2O_3 CO methanation catalyst. *Appl. Catal. A* **2015**, *495*, 104–114. [CrossRef]

25. Rönsch, S.; Köchermann, J.; Schneider, J.; Matthischke, S. Global Reaction Kinetics of CO and CO_2 Methanation for Dynamic Process Modeling. *Chem. Eng. Technol.* **2016**, *39*, 208–218. [CrossRef]

26. Rönsch, S.; Matthischke, S.; Müller, M.; Eichler, P. Dynamische Simulation von Reaktoren zur Festbettmethanisierung—Dynamic Simulation of Fixed-Bed Methanation Reactors. *Chem. Ing. Technol.* **2014**, *86*, 1198–1204. [CrossRef]

27. Eilers, H.; Schaub, G. Fischer-Tropsch-Synthese unter instationären Bedingungen im Suspensionsreaktor: Experimentelle und rechnerische Studien—Fischer-Tropsch Synthesis under Transient Conditions in a Slurry Reactor: Experimental and Mathematical Investigations. *Chem. Ing. Technol.* **2015**, *87*, 837–842. [CrossRef]

28. Newton, M.A. Dynamic adsorbate/reaction induced structural change of supported metal nanoparticles: Heterogeneous catalysis and beyond. *Chem. Soc. Rev.* **2008**, *37*, 2644–2657. [CrossRef] [PubMed]

29. Stötzel, J.; Frahm, R.; Kimmerle, B.; Nachtegaal, M.; Grunwaldt, J.-D. Oscillatory Behavior during the Catalytic Partial Oxidation of Methane: Following Dynamic Structural Changes of Palladium Using the QEXAFS Technique. *J. Phys. Chem. C* **2012**, *116*, 599–609. [CrossRef]

30. Grunwaldt, J.-D.; Molenbroek, A.M.; Topsøe, N.Y.; Topsøe, H.; Clausen, B.S. *In Situ* Investigations of Structural Changes in Cu/ZnO Catalysts. *J. Catal.* **2000**, *194*, 452–460. [CrossRef]

31. Hansen, P.L.; Wagner, J.B.; Helveg, S.; Rostrup-Nielsen, J.R.; Clausen, B.S.; Topsøe, H. Atom-resolved imaging of dynamic shape changes in supported copper nanocrystals. *Science* **2002**, *295*, 2053–2055. [CrossRef] [PubMed]

32. Gänzler, A.M.; Casapu, M.; Boubnov, A.; Müller, O.; Conrad, S.; Lichtenberg, H.; Frahm, R.; Grunwaldt, J.-D. *Operando* spatially and time-resolved X-ray absorption spectroscopy and infrared thermography during oscillatory CO oxidation. *J. Catal.* **2015**, *328*, 216–224. [CrossRef]

33. Eslava, J.L.; Iglesias-Juez, A.; Agostini, G.; Fernández-García, M.; Guerrero-Ruiz, A.; Rodríguez-Ramos, I. Time-Resolved XAS Investigation of the Local Environment and Evolution of Oxidation States of a Fischer–Tropsch Ru–Cs/C Catalyst. *ACS Catal.* **2016**, *6*, 1437–1445. [CrossRef]

34. Mutz, B.; Carvalho, H.W.P.; Kleist, W.; Grunwaldt, J.-D. Dynamic transformation of small Ni particles during methanation of CO_2 under fluctuating reaction conditions monitored by *operando* X-ray absorption spectroscopy. *J. Phys. Conf. Ser.* **2016**, *712*, 012050. [CrossRef]

35. Abbas, Z.; Mezher, T.; Abu-Zahra, M.R.M. Evaluation of CO_2 Purification Requirements and the Selection of Processes for Impurities Deep Removal from the CO_2 Product Stream. *Energy Procedia* **2013**, *37*, 2389–2396. [CrossRef]

36. Köppel, W.; Götz, M.; Graf, F. Biogas upgrading for injection into the gas grid. *Gwf-Gas Erdgas.* **2009**, *150*, 26–35.

37. Burattin, P.; Che, M.; Louis, C. Metal Particle Size in Ni/SiO_2 Materials Prepared by Deposition−Precipitation: Influence of the Nature of the Ni(II) Phase and of Its Interaction with the Support. *J. Phys. Chem. B* **1999**, *103*, 6171–6178. [CrossRef]

38. Bitter, J.H.; van der Lee, M.K.; Slotboom, A.G.T.; van Dillen, A.J.; de Jong, K.P. Synthesis of Highly Loaded Highly Dispersed Nickel on Carbon Nanofibers by Homogeneous Deposition–Precipitation. *Catal. Lett.* **2003**, *89*, 139–142. [CrossRef]

39. Garbarino, G.; Riani, P.; Magistri, L.; Busca, G. A study of the methanation of carbon dioxide on Ni/Al$_2$O$_3$ catalysts at atmospheric pressure. *Int. J. Hydrogen Energy* **2014**, *39*, 11557–11565. [CrossRef]

40. Jung, Y.-S.; Yoon, W.-L.; Lee, T.-W.; Rhee, Y.-W.; Seo, Y.-S. A highly active Ni-Al$_2$O$_3$ catalyst prepared by homogeneous precipitation using urea for internal reforming in a molten carbonate fuel cell (MCFC): Effect of the synthesis temperature. *Int. J. Hydrogen Energy* **2010**, *35*, 11237–11244. [CrossRef]

41. Seo, J.G.; Youn, M.H.; Park, D.R.; Nam, I.; Song, I.K. Hydrogen production by steam reforming of liquefied natural gas (LNG) over Ni–Al$_2$O$_3$ catalysts prepared by a sequential precipitation method: Effect of precipitation agent. *Int. J. Hydrogen Energy* **2009**, *34*, 8053–8060. [CrossRef]

42. Gao, J.; Wang, Y.; Ping, Y.; Hu, D.; Xu, G.; Gu, F.; Su, F. A thermodynamic analysis of methanation reactions of carbon oxides for the production of synthetic natural gas. *RSC Adv.* **2012**, *2*, 2358–2368. [CrossRef]

43. Xu, L.; Wang, F.; Chen, M.; Zhang, J.; Yuan, K.; Wang, L.; Wu, K.; Xu, G.; Chen, W. CO$_2$ methanation over a Ni based ordered mesoporous catalyst for the production of synthetic natural gas. *RSC Adv.* **2016**, *6*, 28489–28499. [CrossRef]

44. Frahm, R. Quick scanning EXAFS: First experiments. *Nucl. Instrum. Methods Phys. Res. Sect. A* **1988**, *270*, 578–581. [CrossRef]

45. Frahm, R. New method for time dependent X-ray absorption studies. *Rev. Sci. Instrum.* **1989**, *60*, 2515–2518. [CrossRef]

46. Nachtegaal, M.; Müller, O.; König, C.; Frahm, R. QEXAFS: Techniques and Scientific Applications for Time-Resolved XAS. In *X-ray Absorption and X-ray Emission Spectroscopy*; van Bokhoven, J.A., Lamberti, C., Eds.; Wiley-VCH: Weinheim, Germany, 2016; pp. 155–183.

47. Doronkin, D.E.; Casapu, M.; Günter, T.; Müller, O.; Frahm, R.; Grunwaldt, J.-D. *Operando* Spatially- and Time-Resolved XAS Study on Zeolite Catalysts for Selective Catalytic Reduction of NO$_X$ by NH$_3$. *J. Phys. Chem. C* **2014**, *118*, 10204–10212. [CrossRef]

48. Abild-Pedersen, F.; Lytken, O.; Engbæk, J.; Nielsen, G.; Chorkendorff, I.; Nørskov, J.K. Methane activation on Ni(111): Effects of poisons and step defects. *Surf. Sci.* **2005**, *590*, 127–137. [CrossRef]

49. Molenbroek, A.M.; Nørskov, J.K.; Clausen, B.S. Structure and Reactivity of Ni−Au Nanoparticle Catalysts. *J. Phys. Chem. B* **2001**, *105*, 5450–5458. [CrossRef]

50. Vang, R.T.; Honkala, K.; Dahl, S.; Vestergaard, E.K.; Schnadt, J.; Laegsgaard, E.; Clausen, B.S.; Norskov, J.K.; Besenbacher, F. Controlling the catalytic bond-breaking selectivity of Ni surfaces by step blocking. *Nat. Mater.* **2005**, *4*, 160–162. [CrossRef] [PubMed]

51. Bengaard, H.S.; Nørskov, J.K.; Sehested, J.; Clausen, B.S.; Nielsen, L.P.; Molenbroek, A.M.; Rostrup-Nielsen, J.R. Steam Reforming and Graphite Formation on Ni Catalysts. *J. Catal.* **2002**, *209*, 365–384. [CrossRef]

52. Kirstein, W.; Petraki, I.; Thieme, F. A study on oxygen adsorption and coadsorption with carbonmonoxide on a stepped nickel surface. *Surf. Sci.* **1995**, *331*, 162–167. [CrossRef]

53. Marwood, M.; Doepper, R.; Renken, A. In-situ surface and gas phase analysis for kinetic studies under transient conditions The catalytic hydrogenation of CO$_2$. *Appl. Catal. A* **1997**, *151*, 223–246. [CrossRef]

54. Pan, Q.; Peng, J.; Sun, T.; Wang, S.; Wang, S. Insight into the reaction route of CO$_2$ methanation: Promotion effect of medium basic sites. *Catal. Commun.* **2014**, *45*, 74–78. [CrossRef]

55. Andersson, M.P.; Abild-Pedersen, F.; Remediakis, I.N.; Bligaard, T.; Jones, G.; Engbæk, J.; Lytken, O.; Horch, S.; Nielsen, J.H.; Sehested, J.; et al. Structure sensitivity of the methanation reaction: H$_2$-induced CO dissociation on nickel surfaces. *J. Catal.* **2008**, *255*, 6–19. [CrossRef]

56. Holloway, P.H. Chemisorption and oxide formation on metals: Oxygen-nickel reaction. *J. Vac. Sci. Technol.* **1981**, *18*, 653–659. [CrossRef]

57. Besenbacher, F.; Nørskov, J.K. Oxygen chemisorption on metal surfaces: General trends for Cu, Ni and Ag. *Prog. Surf. Sci.* **1993**, *44*, 5–66. [CrossRef]

58. Dahl, S.; Logadottir, A.; Egeberg, R.; Larsen, J.; Chorkendorff, I.; Törnqvist, E.; Nørskov, J.K. Role of steps in N$_2$ activation on Ru(0001). *Phys. Rev. Lett.* **1999**, *83*, 1814. [CrossRef]

59. Gracia, F.J.; Bollmann, L.; Wolf, E.E.; Miller, J.T.; Kropf, A.J. In situ FTIR, EXAFS, and activity studies of the effect of crystallite size on silica-supported Pt oxidation catalysts. *J. Catal.* **2003**, *220*, 382–391. [CrossRef]

60. Kale, M.J.; Christopher, P. Utilizing Quantitative *in Situ* FTIR Spectroscopy To Identify Well-Coordinated Pt Atoms as the Active Site for CO Oxidation on Al$_2$O$_3$-Supported Pt Catalysts. *ACS Catal.* **2016**, *6*, 5599–5609. [CrossRef]

61. Boubnov, A.; Gänzler, A.; Conrad, S.; Casapu, M.; Grunwaldt, J.-D. Oscillatory CO Oxidation Over Pt/Al$_2$O$_3$ Catalysts Studied by In situ XAS and DRIFTS. *Top. Catal.* **2013**, *56*, 333–338. [CrossRef]

62. Jacobsen, C.J.H.; Dahl, S.; Hansen, P.L.; Törnqvist, E.; Jensen, L.; Topsøe, H.; Prip, D.V.; Møenshaug, P.B.; Chorkendorff, I. Structure sensitivity of supported ruthenium catalysts for ammonia synthesis. *J. Mol. Catal. A Chem.* **2000**, *163*, 19–26. [CrossRef]

63. Van Hardeveld, R.; van Montfoort, A. The influence of crystallite size on the adsorption of molecular nitrogen on nickel, palladium and platinum. *Surf. Sci.* **1966**, *4*, 396–430. [CrossRef]

64. Mortensen, P.M.; Grunwaldt, J.-D.; Jensen, P.A.; Jensen, A.D. Influence on nickel particle size on the hydrodeoxygenation of phenol over Ni/SiO$_2$. *Catal. Today* **2016**, *259*, 277–284. [CrossRef]

65. Benfield, R.E. Mean coordination numbers and the non-metal-metal transition in clusters. *J. Chem. Soc. Faraday Trans.* **1992**, *88*, 1107–1110. [CrossRef]

66. Geus, J.W.; van Dillen, A.J. Preparation of Supported Catalysts by Deposition–Precipitation. In *Handbook of Heterogeneous Catalysis*; Ertl, G., Knözinger, H., Schüth, F., Weitkamp, J., Eds.; Wiley-VCH: Weinheim, Germany, 2008; pp. 428–467.

67. De Jong, K.P. Deposition Precipitation. In *Synthesis of Solid Catalysts*; de Jong, K.P., Ed.; Wiley-VCH: Weinheim, Germany, 2009; pp. 111–134.

68. Van der Lee, M.K.; van Dillen, J.; Bitter, J.H.; de Jong, K.P. Deposition Precipitation for the Preparation of Carbon Nanofiber Supported Nickel Catalysts. *JACS* **2005**, *127*, 13573–13582. [CrossRef] [PubMed]

69. Bergeret, G.; Gallezot, P. Particle Size and Dispersion Measurements. In *Handbook of Heterogeneous Catalysis*; Ertl, G., Knözinger, H., Schüth, F., Weitkamp, J., Eds.; Wiley-VCH: Weinheim, Germany, 2008; pp. 738–765.

70. Müller, O.; Nachtegaal, M.; Just, J.; Lutzenkirchen-Hecht, D.; Frahm, R. Quick-EXAFS setup at the SuperXAS beamline for *in situ* X-ray absorption spectroscopy with 10 ms time resolution. *J. Synchrotron. Radiat.* **2016**, *23*, 260–266. [CrossRef] [PubMed]

71. Müller, O.; Lützenkirchen-Hecht, D.; Frahm, R. Quick scanning monochromator for millisecond *in situ* and in *operando* X-ray absorption spectroscopy. *Rev. Sci. Instrum.* **2015**, *86*, 093905. [CrossRef] [PubMed]

72. Grunwaldt, J.-D.; van Vegten, N.; Baiker, A. Insight into the structure of supported palladium catalysts during the total oxidation of methane. *Chem. Commun.* **2007**, 4635–4637. [CrossRef] [PubMed]

73. Grunwaldt, J.-D.; Caravati, M.; Hannemann, S.; Baiker, A. X-ray absorption spectroscopy under reaction conditions: Suitability of different reaction cells for combined catalyst characterization and time-resolved studies. *Phys. Chem. Chem. Phys.* **2004**, *6*, 3037–3047. [CrossRef]

74. Ravel, B.; Newville, M. Athena, Artemis, Hephaestus: Data analysis for X-ray absorption spectroscopy using IFEFFIT. *J. Synchrotron. Radiat.* **2005**, *12*, 537–541. [CrossRef] [PubMed]

75. Rehr, J.J.; Albers, R. Theoretical approaches to X-ray absorption fine structure. *Rev. Mod. Phys.* **2000**, *72*, 621. [CrossRef]

MDPI

St. Alban-Anlage 66

4052 Basel

Switzerland

Tel. +41 61 683 77 34

Fax +41 61 302 89 18

www.mdpi.com

Catalysts Editorial Office

E-mail: catalysts@mdpi.com

www.mdpi.com/journal/catalysts

www.ingramcontent.com/pod-product-compliance
Lightning Source LLC
Chambersburg PA
CBHW051911210326
41597CB00033B/6105